Explorers

MAGILL'S CHOICE

EXPLORERS

Volume 2
Ibn Battutah —
Fanny Bullock Workman
309 – 590

from
THE EDITORS OF SALEM PRESS

SALEM PRESS, INC.
Pasadena, California   Englewood Cliffs, New Jersey
Ypsilanti District Library
5577 Whittaker Rd.
Ypsilanti, MI 48197-9752

Copyright © 1998, by SALEM PRESS, INC.
All rights in this book are reserved. No part of this work may be used or reproduced in any manner whatsoever or transmitted in any form or by any means, electronic or mechanical, including photocopy, recording, or any information storage and retrieval system, without written permission from the copyright owner except in the case of brief quotations embodied in critical articles and reviews. For information address the publisher, Salem Press, Inc., P.O. Box 50062, Pasadena, California 91115.

Essays originally appeared in the following series, all edited by Frank N. Magill: *Great Lives from History: Ancient and Medieval Series*, 1988; *Great Lives from History: Renaissance to 1900 Series*, 1989; *Great Lives from History: British and Commonwealth Series*, 1987; *Great Lives from History: American Series*, 1987; and *Great Lives from History: American Women Series*, 1995. New material has been added.

∞ The paper used in these volumes conforms to the American National Standard for Permanence of Paper for Printed Library Materials, Z39.48-1992.

Library of Congress Cataloging-in-Publication Data
Explorers
      p.   cm. — (Magill's choice)
"From the editors of Salem Press."
Includes bibliographical references and index.
  ISBN 0-89356-970-4 (set : alk. paper). — ISBN 0-89356-971-2 (v. 1 : alk. paper). — ISBN 0-89356-976-3 (v. 2 : alk. paper).
  1. Explorers—Biography—Dictionaries. I. Series.
G200.E874   1998
910'.92'2—dc21
[B]   98-35398
                                                                         CIP

Second Printing

PRINTED IN THE UNITED STATES OF AMERICA

Contents — Volume 2

Ibn Battutah 309

La Salle, Sieur de 315
Leif Eriksson 322
Lewis, Meriwether *and* William Clark 329
Lindbergh, Charles A. 338
Livingstone, David 349

Mackenzie, Sir Alexander 356
Magellan, Ferdinand 363
Mas'udi, al- 372
Menéndez de Avilés, Pedro 377
Muir, John 383

Nansen, Fridtjof 389

Oglethorpe, James Edward 396

Park, Mungo 405
Peary, Robert Edwin 412
Penn, William 422
Pike, Zebulon Montgomery 430
Pizarro, Francisco 437
Polo, Marco 443
Ponce de León, Juan 451

Powell, John Wesley 457
Pytheas 465

Ralegh, Sir Walter 471
Rasmussen, Knud Johan Victor 480
Ride, Sally 487

Sacagawea 494
Smith, Jedediah Strong 502
Smith, John 509
Soto, Hernando de 515
Speke, John Hanning 524
Stanley, Henry Morton 530

Vespucci, Amerigo 539

Wilkes, Charles 547
Wilkins, Sir George Hubert 553
William of Rubrouck 561
Winthrop, John 567
Workman, Fanny Bullock 574

Time Line of Explorers 581

Index 593

EXPLORERS

IBN BATTUTAH

Born: 1304; Tangier, Morocco
Died: c. 1377; Morocco

Driven by an exceptional wanderlust, Ibn Battutah became the greatest Muslim traveler. His peregrinations through India, Russia, China, the East Indies, North Africa, and the Near and Middle East were recorded in the most famous of all Islamic travelogs, the Rihlah.

Early Life

Abu 'Abd Allah Muhammad ibn 'Abd Allah al-Lawati al-Tanji, who came to be known as Ibn Battutah, was born in Tangier, Morocco. His family had a long tradition of serving as religious judges, and his educational training prepared him for such a career. Well before he undertook the first of his many journeys at the age of twenty-one, Ibn Battutah had studied Islamic theology and law in Tangier. His first journey, which commenced in June of 1325, took the form of a pilgrimage to Mecca, but it had much broader ramifications. In a fashion reminiscent of the grand tours which Europeans would undertake in later centuries to finish their education, his trip to Mecca supplied Ibn Battutah with diverse opportunities.

Ever the astute observer and inquisitive intellectual, Ibn Battutah talked and studied with scholars he encountered as he made an unhurried progress eastward. Evidently his family was an affluent one, with extensive connections throughout the Islamic world. Their belief—a view seemingly shared by Ibn Battutah—was that his experiences would be ideally suited to the duties of a magistrate, which he was expected to assume upon his return. His purse, personality, and contacts with powerful officials readily opened many doors for the young man, and his winning ways, together with a considerable degree of curiosity he aroused as an individual from the outer geographic reaches of Islam, stood him in good stead. He was greeted with hospitality wherever he went.

Ibn Battutah's initial journey was a momentous one in a number of ways. He traveled to Cairo, probably the greatest city of the time, making stops, en route from Tangier, in most of the major ports of the southern Mediterranean. He reveled in the time he spent in that renowned intellectual center of ancient times, Alexandria. His experiences in Egypt aroused in him an insatiable wanderlust. He reached Damascus in August, 1326, and it was there that he took his first wife. After a brief courtship and honeymoon, he joined a caravan of pilgrims wending their way to Mecca.

The pilgrims' travels were arduous in the extreme. After passing through what is now Jordan and Syria, the faithful rested for several days at Al-Karak, for the ordeal of a desert crossing lay before them. They managed the crossing of the Wadi al-Ukhaydir—Ibn Battutah characterized it as the valley of hell—by moving at night until they reached the Al-'Ula oasis. Thence they moved onward to the holy city of Medina and then progressed to Mecca. As was the case with all of his peregrinations, Ibn Battutah produced vivid accounts of his experiences in and impressions of Mecca. He departed from Mecca sometime in November of 1326, now a *haji* (one who has made the pilgrimage to Mecca) as well as a man determined to see much more of the world.

Life's Work

The essence of Ibn Battutah's achievement lies in his wide-ranging travels. Yet there was much more to his repeated journeys than merely visiting strange and faraway places. The reader of *Tuhfat al-nuzzar fi ghara'ib al-amsar wa-'aja'ib al-asfar* (1357-1358; *The Travels of Ibn-Battuta*, 1958-1980, best known as *Rihlah*) becomes immediately aware of Ibn Battutah's keen eye for detail; clearly, he was a man of rare curiosity and intellect. Anything and everything interested him. From unusual religious beliefs to the economies of the regions through which he trekked, from methods of dress to the basics of diet, he noted the varied aspects of the lifestyles which he encountered. Indeed, what Ibn Battutah saw and reported is at least as important as where he went.

Returning with his two wives to Tangier, Ibn Battutah tarried only a short time before succumbing again to his desire to travel. His next major undertaking was a second *hajj*, and this time he gave himself ample time to sense and savor all Mecca had to offer. His visit lasted for some three years. This must have been quite expensive, but evidently the family fortunes were ample to support such extended periods of travel.

Next, Ibn Battutah decided to sail down the Red Sea to the renowned trading center of Aden. He provides a singular description of the strategically located seaport and the way in which it depended on great cisterns for its water supply. After some time in Aden, he continued southward along the African coast and visited the important trade centers of Kilwa and Mombasa. On his return journey, he stopped in the major cities of Oman and Hormuz. In fact, in a fashion which was to become characteristic of all his endeavors, he attempted to visit every reachable site of major significance.

After touring the Gulf of Aden and its environs, he traversed the considerable breadth of Arabia while making the *hajj* for a third time. A trip across the Red Sea followed, with a risky, demanding journey to Syene (modern Aswan) and thence via the Nile to Cairo. By this juncture, Arabia, Africa's Mediterranean coast, and the lower reaches of the Nile had become familiar territory, and, not surprisingly for a man of his inclinations, Ibn Battutah began to look farther afield.

He passed through the various Turkish states in Asia Minor, crossed the Black Sea to reach Kaffa (modern Feodosiya, the first Christian city he had seen), and moved northeastward into Kipchak. This Russian region was then under the control of Khan Muhammad Özbeg, and Ibn Battutah joined his peripatetic camp. In this way he was able to visit the outer reaches of Mongolia, where he marveled at the brevity of summer nights.

Upon leaving the khan's camp, Ibn Battutah linked his travel fortunes to a Byzantine princess, whom he accompanied to Constantinople. There, in what he considered one of the most important moments of his career, he enjoyed an interview with Emperor Andronikos III. From the emperor's court he journeyed eastward, crossing the steppes of southern Asia en route to Kabul and thence over the Hindu Kush. In September of 1333, he reached the Indus River. He had, in eight years of traveling, made three pilgrimages to Mecca, seen most of the southern and eastern Mediterranean, floated on the Nile, braved the desert sands of Arabia, and penetrated deep into Russia. It is at this juncture that he ends his first narrative, and by any standard of measurement his achievements had been considerable. Still, though he was probably the most traveled man of his time, the clarion call to adventure drew him as strongly as ever.

He wandered throughout the Sind and eventually moved on to Delhi at the invitation of the ruler, Muhammad ibn Tughluq. This capricious, bloodthirsty monarch was a bit too much even for Ibn Battutah's eclectic tastes: "No day did his palace gate fail to witness the elevation of some abject to affluence and the torture and murder of some living soul." Yet somehow he managed to get along with this extraordinary ruler, and he became *Qadi* (judge) of Delhi at a very high salary. He served in this capacity for the next eight years. Yet, far from prospering, he fell into considerable debt. Scholars have ascribed this to his extravagance, and his living beyond his means may have figured prominently in his eventual decline into disfavor.

Resilient soul that he was, however, Ibn Battutah turned potential ruin into what to him must have been a glorious assignment. He was chosen to head a delegation which was paying a visit to the last Mongol emperor of China. Leaving in 1342, the group made its way to Calcutta en route to China. Here fate intervened, however, in the form of a shipwreck that completely destroyed the junk on which he and the other envoys were to travel. This was a disaster of the first magnitude, for Ibn Battutah lost not only his personal possessions but also the lavish gifts he had been delegated to carry to China.

Accordingly, Ibn Battutah remained in the region, visiting various cities on India's western coast and also the Maldive Islands, where he rose to prominence as a judge and added four wives to his harem. Yet he did not tarry overlong, for August, 1344, found him leaving the Maldives for Ceylon (modern Sri Lanka). Further adventures followed, and eventually he reached Java after having stopped briefly in Burma. From Java he finally made his way to China, where he visited Amoy, Canton, and Peking, among other major sites. Returning westward, he revisited Sumatra, Malabar, Oman, Persia, and a host of other locations. Upon reaching Damascus, he learned of his father's death some fifteen years earlier. It was the first news of home he had had in that many years.

It was also during this period that Ibn Battutah witnessed, at first hand, the ravages of the bubonic plague (sometimes called the Black Death). His graphic reports on what he saw in Damascus, where more than two thousand unfortunate souls died in a single day, is one of the finest surviving accounts of the plague. Perhaps thereby reminded of his mortality, he then revisited Jerusalem and Cairo in the process of making a fourth *hajj*. Finally, having been away from home almost constantly for more than twenty-four years, he returned to Morocco on November 8, 1349.

Even then, his travels were not over. After spending a relatively short time in Tangier, he made his way to Spain and toured Andalusia. His final major journey was into central Africa. Journeying from oasis to oasis across the Sahara, he reached the fabled desert entrepôt of Tombouctou, where the mighty Niger River (which he wrongly called the Nile) begins its great westward sweep to the ocean. At this point, his king called him home, thereby bringing an end to nearly thirty years of travel encompassing an estimated seventy-five thousand miles. His final years were more settled; he was in his early seventies when he died in his native Morocco.

Impact

With the possible exception of the voyages of Marco POLO, there is nothing prior to the European Renaissance to compare with the nature and extent of Ibn Battutah's travels. He single-handedly made the world a smaller place, and thanks to his remarkable accounts modern knowledge of much of Asia during the fourteenth century is considerably richer than otherwise would have been the case. As the chronicler Muhammad ibn Juzayy, who recounted Ibn Battutah's travels by royal decree, stated: "This Shaykh is the traveller of our age; and he who should call him the traveller of the whole body of Islam would not exceed the truth."

Bibliography

Cooley, William D. *The Negroland of the Arabs Examined and Explained*. London: J. Arrowsmith, 1841. Although there are some problems with Cooley's transcriptions from the Arabic, this is a useful early English account of that portion of Ibn Battutah's travels devoted to the Sahara and Niger regions.

Gibb, H. A. R., ed. *Travels in Asia and Africa, 1325-1354*. New York: A. M. Kelley, 1969. A reprint of the 1929 edition in the Argonaut series, this work is a convenient, accurate account of the highlights of Ibn Battutah's career. Gibb's introduction and notes offer useful historical background.

Hamilton, Paul. "Seas of Sand." In *Exploring Africa and Asia*, by Nathalie Ettinger, Elspeth J. Huxley, and Paul Hamilton. Garden City, N.Y.: Doubleday Publishing Co., 1973. The section entitled "The Traveler of Islam" constitutes a useful, accessible account of Ibn Battutah's first journey to Mecca. This volume is part of the Encyclopedia of Discovery and Exploration series. Illustrated.

Ibn Battutah. *The Travels of Ibn-Battuta*. Translated and edited by H. A. R. Gibb. 4 vols. Cambridge: Cambridge University Press, 1958-1980. This careful, amply annotated translation is by far the most important English-language source of information on the man and his milieu. Part of the Hakluyt Society's series on early explorers.

Tucker, William. "Ibn-Battuta, Abu Abd-Allah Muhammad." In *The Discoverers: An Encyclopedia of Exploration*, edited by Helen Delpar. New York: McGraw-Hill Book Co., 1980. A succinct, useful summation of the high points of Ibn Battutah's career. This volume includes short biographies of many other explorers as well. Includes bibliographies and an index.

James A. Casada

SIEUR DE LA SALLE

Born: November 21, 1643; Rouen, France
Died: March 19, 1687; on the Brazos River, Texas

La Salle was the first European to traverse fully the Mississippi River. He exited into the Gulf of Mexico, where he later attempted unsuccessfully to found a French colony on the Texas coast.

Early Life

René-Robert Cavelier was born November 21, 1643, at Rouen, in Normandy. He preferred to use as his name the noble title Sieur de La Salle. His father, Jean Cavelier, was a wealthy landowner. La Salle's mother, née Catherine Geest, came from a family of wholesale merchants. The parents wanted René-Robert and his older brother, Jean, to take priestly vows in the Catholic Church. Jean joined the Order of Saint Sulpice and went to French Canada. Young René-Robert entered the Jesuit College of Rouen several years later, at the age of nine.

At Rouen, the youth showed an aptitude for mathematics and philosophy. He thus proved to be a capable student in his academic work. René-Robert, however, had problems with the Jesuit teachers. His large physical size and athletic prowess, coupled with a lusty desire for adventure and excitement, did not suit him for a life of prayer and scholarship. Although no specific physical description of him survives, La Salle seems to have been a handsome and personable youth who made friends easily. At times, however, young La Salle proved to be moody, hot-tempered, and stubborn. His rebellious spirit and strong sense of independence made it difficult for him to succeed in the Jesuit brotherhood. This calling demanded introspection, moderation, tolerance, and austerity. Nevertheless, the youth completed his education at Rouen, although he refused to take full vows as a Jesuit brother upon reaching adulthood.

LA SALLE EXPEDITION

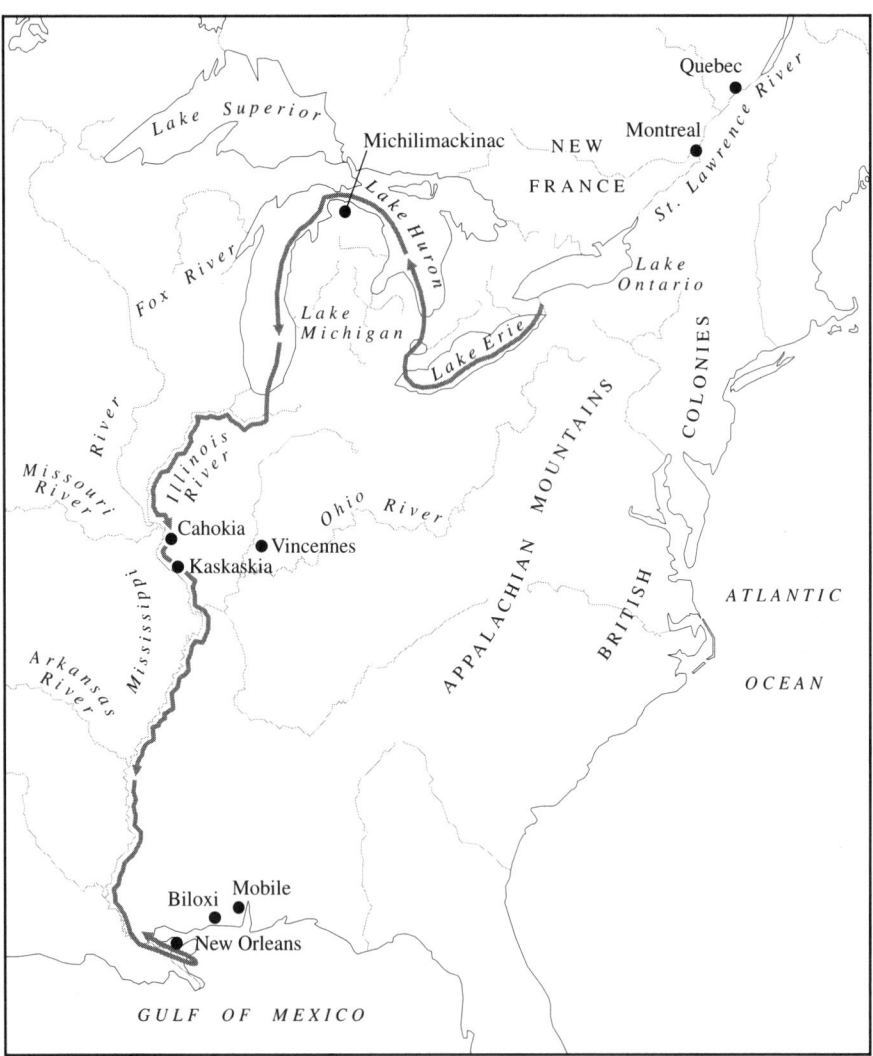

Instead, young Cavelier renounced his novitiate vows in 1667 and decided to emigrate to New France. He was penniless upon leaving the Jesuits because his father had died while he was a novitiate in the order. Under French law, René-Robert had thus been ineligible to inherit the family property. Canada might therefore provide the young man a chance to earn fortune and fame. In addition, La Salle already had connections in the colony through his brother Jean at Montreal. Also, his uncle had been a member of the Hundred Association of New France and a heavy investor in French development of Canada.

La Salle used these family connections to secure a seigniory (a large landed estate) along the St. Lawrence River. There, he began living the life of a gentleman planter. He traveled in the best social circles of the colony, meeting the rich and powerful of New France. La Salle soon became fascinated with the Indians of Canada and learned some of their major languages. He realized that fur trading with the Indians would provide for him the fastest route to wealth and riches. This desire to enter the fur trade caused La Salle to sell his seigniory and move to the frontier.

Life's Work
La Salle joined an expedition sent by the Sulpicians in 1669 to found new missions and trade in furs along the western Great Lakes. He visited the Ohio River Valley and familiarized himself with much of the Great Lakes region. This expedition established La Salle as a successful explorer who had the potential for expanding the French fur trade into new areas of North America. His activities on the western frontier caught the attention of New France's governor, Louis de Baude, the Comte de Frontenac. The governor was anxious to enrich himself personally in the fur trade. He recognized that a partnership with La Salle would provide a means of doing so. The governor dispatched La Salle to France in 1674 to obtain for them a royal fur-trading monopoly in the Ontario region of the St. Lawrence River. This the two partners soon secured from the French king. From his base at Fort Frontenac on the eastern end of Lake Ontario, La Salle spent the next three years exploring and trading with the Indians in the upper Great Lakes region.

All the while, La Salle dreamed of greater triumphs. He had heard rumors of the mighty unexplored river to the west, which the Indians called "MessiSipi." He had also talked with Louis Jolliet, another Frenchman who had earlier visited its upper reaches. La Salle decided to secure a grant from the king permitting him to descend the river to its mouth, explore its course in the process, and, in so doing, obtain a fur-trading monopoly with the Indians along its banks. In 1677, he returned to France in the company of his faithful lieutenant Henry de Tonty and secured such a concession from the king. By the summer of 1678, La Salle was back in Canada, busily engaged in organizing an expedition for this purpose.

Events, however, moved slowly. It took more than a year to raise the necessary money and secure all the supplies. Further delays came when La Salle suffered various financial reverses. Finally, early in 1682, he and his expedition began their descent of the Mississippi, reaching the Gulf of Mexico on April 9. In formal ceremonies held on that date near the mouth of the river, La Salle laid claim for France to all lands which the river drained. He named the region "Louisiana," in honor of the French king. La Salle and his men then returned whence they had come, making the laborious journey back up the river to Canada.

During La Salle's absence on the Mississippi expedition, Frontenac left the governor's office, and Antoine Lefebre, the Sieur de la Barre, assumed the position. He was one of La Salle's enemies in the competitive fur trade of French Canada. The new governor removed La Salle from his position in absentia and accused him of various minor crimes. La Salle returned to France to clear his name. The king, pleased with the explorer's accomplishments, restored his monopoly and trading rights in the Mississippi Valley. La Salle thereupon set about organizing a major colonizing expedition which would found a French settlement on the Gulf of Mexico at the mouth of the Mississippi. This expedition left France in July of 1684. Its voyage across the Atlantic was not auspicious. Food and water were in short supply, while La Salle quarreled incessantly with the Sieur de Beaujeu, the naval officer who commanded the ships of the expedition.

Morale was low and illness had ravaged the colonists by the time they stopped at French Hispañola, their stepping stone to the Gulf of Mexico. La Salle, suffering from a fever, was indecisive in his leadership while Captain Beaujeu refused to cooperate with the explorer. Under these circumstances, they continued their journey. The reconnaissance to find the mouth of the Mississippi did not go well. La Salle missed the river entirely, instead leading his expedition almost due west to Matagorda Bay on the shores of Texas.

Nevertheless, La Salle decided to establish his colony in this uncharted territory and use it as a base to search for the Mississippi. He had a small fort constructed while the colonists built modest huts in which to live. Conditions in the colony were harsh and forbidding, made worse by La Salle's quarrel with Captain Beaujeu. Acrimony with Beaujeu became so bitter that the naval commander sailed home, leaving the colonists on their own. Moreover, La Salle was not clear about his intentions. At times he talked of harassing the Spanish to the south in Mexico, while on other occasions he maintained that finding the Mississippi was his chief objective.

During 1686, La Salle led several exploring parties into various parts of the surrounding region. The only result of these journeys was increased dissatisfaction among the colonists. In January, 1687, La Salle and a small band of men departed on foot for the Mississippi, leaving behind most of the colonists at Fort St. Louis. By March, several members of this traveling party had become frustrated with La Salle's authoritarian style of command. These individuals plotted to assassinate La Salle by ambush, which they did on March 19, 1687.

The death of La Salle ended the French colony at Fort St. Louis, which did not long survive his passing. Within months, some of the colonists made their way back to Canada via the Mississippi (which they eventually found), while others died at the hands of hostile Indians who attacked the settlement. A handful of colonists, including several children, lived with friendly Indians along the Texas coast. Spanish captain Alonso de León, whom the authorities in Mexico City had sent to destroy the French settlement, found the few remaining survivors when he arrived at Fort St. Louis in 1689.

Impact

La Salle's greatest triumph was his exploration of the Mississippi River in 1682, which established claim to Louisiana for France. This acquisition more than doubled the territory held by the French king in North America. Yet the spirit of independence and single-mindedness which permitted La Salle to excel as an explorer made him poorly suited to be a colony builder. He sometimes treated his subordinates dogmatically and imperiously. He could be moody and withdrawn to the extent that one modern biographer, E. B. Osler, maintains that La Salle was a manic-depressive personality type. Moreover, the explorer was a poor financial manager and spent most of his career deeply in debt.

Nevertheless, the Sieur de La Salle ranks as one of history's best-known explorers. His accomplishments include a number of "firsts" which certainly justify this reputation: the first European to travel down the Mississippi to the Gulf; the first person to advocate the founding of a major city at the mouth of that river; and the first colonizer to attempt a settlement on the western coast of the Gulf of Mexico. It is because of La Salle that the French fleur-de-lis is one of the six flags which have flown over Texas during its history.

Bibliography

Caruso, John Anthony. *The Mississippi Valley Frontier: The Age of French Exploration and Settlement*. Indianapolis: Bobbs-Merrill Co., 1966. Standard scholarly survey of expansion of French Canada into Louisiana. Examines La Salle and the colonizing activities which came after him in the eighteenth century.

Cox, Issac Joslin. *The Journeys of René-Robert Cavelier, Sieur de La Salle*. 2 vols. New York: A. S. Barnes and Co., 1905-1906. Lengthy collection of documents, personal memoirs, and contemporary reports dealing with La Salle's career. All in English translation.

Delanglez, Jean. *Some La Salle Journeys*. Chicago: Institute of Jesuit History, 1938. Stresses La Salle's geographic discoveries and details his explorations by use of cartographic evidence.

Galloway, Patricia, ed. *La Salle and His Legacy: Frenchmen and Indians in the Lower Mississippi Valley*. Jackson: University of Mississippi Press, 1982. Series of historical papers on La Salle presented at the 1982 meeting of the Mississippi State Historical Society, in celebration of the three hundredth anniversary of La Salle's voyage.

Joutel, Henri. *Joutel's Journal of La Salle's Last Voyage*. London: A. Bell, 1714. Reprint. Edited by Melville B. Anderson. Chicago:

The Caxton Club, 1896. Joutel accompanied La Salle on the Texas expedition and was among the survivors.

Muhlstein, Anka. *La Salle: Explorer of the North American Frontier*. New York: Little, Brown and Co., 1994. Translated from the French *Cavelier de La Salle: Ou, L'Homme qui offrit l'Amerique à Louis XIV* by Willard Wood. Includes maps and bibliography.

Osler, E. B. *La Salle*. Don Mills, Ontario: Longmans Canada, 1967. The best modern biography. Full of detail, it takes the position that La Salle was sometimes mentally unstable, thereby explaining his erratic leadership style.

Parkman, Francis. *La Salle and the Discovery of the Great West*. Boston: Little, Brown and Co., 1879. Parkman ranks as one of the greatest nineteenth century narrative historians. The study, in spite of minor inaccuracies, is a most literary and readable work dealing with La Salle.

Weddle, Robert S. *The Spanish Sea*. College Station: Texas A&M University Press, 1985. A lengthy examination of European efforts to explore the Gulf of Mexico in the sixteenth and the seventeenth centuries. Based on extensive manuscript research in European archives.

_____. *Wilderness Manhunt: The Spanish Search for La Salle*. Austin: University of Texas Press, 1973. Examines the La Salle colony in Texas from the Spanish viewpoint, as a threat to Spain's control of the Gulf of Mexico. Contains much detail about Spanish efforts to locate and destroy La Salle's ill-fated colony.

Light Townsend Cummins

LEIF ERIKSSON

Born: c. 970; Iceland, possibly in Haukadal
Died: c. 1035; probably near Julianehaab, Greenland

Though probably not the first European to sight America, Leif made the first deliberate exploration of the North American continent and provided the main stimulus for later, unsuccessful attempts at permanent settlement.

Early Life

Very little is known of Leif's early life. He was the son of Erik the Red and his wife, Thjodhild, and seems to have had two brothers, Thorvald and Thorstein, and one sister, Freydis. His father's career, however, is well-known. Erik was born in Norway but was forced to flee as a result of "some killings." He settled first at Drangar in Iceland but then moved to Haukadal. At that time, though land was still readily available in Iceland, the country had been known for more than a century and intensively settled for perhaps eighty years; there were many powerful and well-established families in all the areas where Erik attempted to settle.

In Haukadal, he became involved in several conflicts, killing at least two of his neighbors, Eyjolf "the Sow" and Hrafn "the Dueler." He was driven out, tried to make his home elsewhere, killed another neighbor in an argument over timber, and was then—not unreasonably—outlawed together with his family.

Erik then made the momentous decision to try to find an unsettled land. Seafarers blown off course had reported land to the west of Iceland, and in 982 Erik sailed, together with his family, to find it. He landed in Greenland near what is now Julianehaab and spent three years exploring the country. In 985 he returned to Iceland and in 986 set sail again with twenty-five ships to found a permanent settlement in Greenland. Only fourteen of the ships arrived, with perhaps four hundred people, but this landing formed the basis for the later colonization of the eastern, middle, and western settlements of Greenland, which lasted until changing climate and Eskimo hostility exterminated the colonies, probably in the early 1500's.

Nevertheless, this colonizing move had transformed Erik from a hunted outlaw in a land severely afflicted by famine to the undisputed head of a new nation, the patriarch of a land with reasonable grazing (in the more temperate climate of the late tenth century) and unparalleled hunting, trapping, and fishing opportunities. It seems reasonable to suppose that the total change of life-style also made an impression on his children, who may have wondered if they too could not become great men or great women by similar daring seamanship.

Archive Photos

Life's Work
To reconstruct Leif's life, two Icelandic sagas are indispensable: *Groehnlendinga saga* (c. 1390; *The Greenlanders' Saga*, 1893) and *Eiríks saga rauda* (c. 1263; *The Saga of Erik the Red*, 1841), the latter existing in two different versions. These sagas do not tell quite the same tale, but reasons for their deviations can often be seen. According to *The Greenlanders' Saga*, which was composed much earlier than its late fourteenth century transcription date, America was originally sighted not by Leif but by one Bjarni Herjolfsson, who had been blown off course on his way to Greenland. Bjarni refused,

however, to land at any of the three places he sighted (to the disgust of his crew) and finally made his way to his father's farm, located about fifty miles from the farm Erik and Leif had established at Brattahlith. Bjarni's sightings caused much discussion, and some time later, probably around the year 1000, Leif came to him and bought his ship—presumably thinking that if the ship had reached this strange destination once, it could do so again. Leif hoped to get his father, now a man of fifty or more, to lead the expedition, because of his famous good luck, but on his way to the ship Erik fell off his horse, hurt himself, and refused to go any farther. It was not his fate to discover more new lands, he said. He would leave that to his son.

The Greenlanders' Saga then relates that Leif and his men came in succession to countries they called Helluland (flatstone land) and Markland (forest land), finally arriving at a place where they stayed for the winter and by which they were much impressed. It had sweet dew, no winter frost, outdoor grazing for cattle all year, and sun visible as late as midafternoon even in midwinter—very different from the short midwinter days of Greenland or Iceland. Finally, an old attendant of Leif, the German Tyrkir, was found one day almost incoherent with delight: He had found wild grapes, from which the land was given the name of Vinland, or Wineland. Leif and his men loaded a cargo of grapes and timber—the latter in very short supply in the treeless northern islands—and went home. On their way, they sighted and rescued a wrecked ship's crew, again men who had been blown off course.

It seems likely that the story cited above is close to what really happened. In later years, however, the rather haphazard nature of the expedition was considered insufficiently inspiring, and *The Saga of Erik the Red* added a rather pointless tale of a love affair between Leif and a Hebridean lady named Thorgunna, which left him with a son, Thorgils, and a tale of how Leif went to Norway to the court of Olaf I Tryggvason, the missionary king, there to be converted to Christianity and sent back to preach the new religion in Greenland. On his return, says this saga, Leif was blown off course, sighted a land with wheat and vines on it, rescued a ship's crew, and finally arrived in Greenland to preach the faith. One can see that in this story the stubborn, unenterprising Bjarni Herjolfsson has vanished; Leif has been given entire credit for discovering Vinland, and the whole story has become vaguely tied to the advantages of Christianity. Very little is said about geography, however, and it is not at all clear how Leif had time to go to Norway, be converted, go to Vinland, explore it, and get back to Greenland, all in one short northern summer. Almost certainly the tale of Leif and the conversion of Greenland is a later addition. It is not known when Greenland was converted, but it probably occurred after the conversion of Iceland in 1000. Leif probably died a Christian, but he was likely still a pagan at the time of his landing in America.

The stories after the discovery deviate even further, but one can make out some consistent elements. Leif could and did tell people how to reach his winter settlement at Leifsbuthir in Vinland—a "booth" being the Norse term for a temporary hut. He had in fact left a house of sorts there and was prepared to lend it to people, especially family members, but not to give it away; he seems to have felt the need to keep a claim of sorts on the country. Yet later visits were not successful. Thorvald, Leif's brother, was killed by an arrow. *The Greenlanders' Saga* says it was shot by a "Skraeling"; *The Saga of Erik the Red*, with its usual attempt to improve a story, by a "uniped." Nevertheless, all sources agree that the native inhabitants of the country, called contemptuously Skraelings, or "wretches, punies," by the Norsemen, became increasingly hostile after early attempts at trade and in the end forced the Norsemen out. Another would-be colonizer, Thorfinn Karlsefni, who had married the widow of Thorstein (Leif's other brother), tried to settle near Leifsbuthir but was also compelled to leave. Finally, an expedition which included Leif's sister Freydis ended in mass murder when Freydis provoked and wiped out the crew of her companion ship, herself killing five women with an ax when no man would do it. She tried to hide the matter from her brother, but when the party returned to Greenland, Leif discovered the truth and, though reluctant to punish her himself, made the killings known. The visits to Vinland had proved unlucky, and Leif and his family attempted to settle there no more—as far as is known from the sagas.

Impact

In a sense, Leif Eriksson did very little to earn his later reputation, or even his Norse nickname, Leif "the Lucky." He did not "discover" America; Bjarni Herjolfsson did. He did not try to colonize it; if anyone can be given credit for that, it should be Thorfinn Karlsefni. What Leif did was to explore the Labrador-Newfoundland-Nova Scotia coast and to publicize his exploration. In this at least he was a master, and stories rapidly became attached to his voyage: The turning back of his father, the discovery of the grapes, the use of a pair of trained Scottish "runners" given to him by King Olaf to scout large areas of land quickly and cheaply without the bother of shipping horses. Some of these stories are probably in essence true. Others were attracted to the saga by the interest taken in these western discoveries.

Furthermore, Vinland may have been visited more often than the sagas state. In the 1950's, Helge Ingstad made a careful search of the Newfoundland coast for relics of Norse settlement, trying to reconcile the geography of the area with such carefully described places in the sagas as Furdustrandir, the Wonder Beaches (perhaps the long expanse of sandy beaches south of Hamilton Inlet in southeast Labrador). In the end, Ingstad found what are claimed to be clear signs of Norse building at L'Anse-aux-Meadows in Newfoundland. Whether these are Leifsbuthir or some later establishment cannot be determined.

Finally, Vinland remained marked on maps drawn in Iceland at least up to 1590, a century after Christopher COLUMBUS. At any time in the fourteenth or fifteenth centuries any Englishman, Italian, or Portuguese who bothered to ask an Icelander about northern geography would probably have been told that there was a large, fertile country west of Greenland. Whether anyone did ask is not known. Yet it cannot be ruled out that, for example, Columbus had sailed "beyond Thule," or Iceland, and in his 1492 expedition was encouraged by dim memories of the voyage of Leif the Lucky.

Bibliography

Gad, Finn. *The History of Greenland*. Vol. 1, *Earliest Times to 1700*. Translated by Ernst Dupont. London: Hurst and Co., 1970. A stirring account of the history of this doomed colony. Gad closely follows the Christianized version of Leif's adventures, perhaps wrongly, but the author has interesting information on the nature of Erik's and Leif's farm at Brattahlith as revealed by archaeology.

Gathorne-Hardy, G. M. *The Norse Discoverers of America*. Oxford: Clarendon Press, 1921. A conflated account of the different saga versions, making for an easier story to read but also rather obscuring the genuine discrepancies. There is a sensible discussion of textual problems and of such issues as the nature of the Skraelings.

Haugen, Einar, trans. *Voyages to Vinland*. New York: Alfred A. Knopf, 1942. This work is a translation of the various sources together with a long account of the evidence as Haugen sees it. Though largely superseded by the work by Jones, below, this book remains attractive for its clear style, and valuable because of Haugen's own status as a scholar of Norse.

Ingstad, Helge. *Westward to Vinland*. Translated by Erik J. Friis. New York: St. Martin's Press, 1969. This account gives the story of what is often regarded as the only convincing Norse archaeological site in the New World, at L'Anse-aux-Meadows in Newfoundland. Doubt is cast on this by Farley Mowat, but if even the bronze pin and the spindle whorl found are incontrovertibly Norse, then it must be accepted that this is a Norse site, possibly Leif's, possibly from an unrecorded subsequent expedition.

Jones, Gwyn. *The Norse Atlantic Saga: Being the Norse Voyages of Discovery and Settlement to Iceland, Greenland, and America*. London: Oxford University Press, 1964. Professor Jones traces the story of the Norse discoveries of Iceland, Greenland, and Vinland, with interesting details on seamanship and on the successive and half-planned nature of Norse exploration. The book includes translations of both *The Greenlanders' Saga* and *The Saga of Erik the Red* as well as translocations of minor tales, including Eskimo ones. The most useful single work on this period and milieu.

Magnusson, Magnus, and Hermann Palsson, trans. *The Vinland Sagas*. New York: New York University Press, 1965. This work gives fresh and clear translations of *The Greenlanders' Saga* and

The Saga of Erik the Red, using as its basis for the latter the fuller if later *Skalholtsbok* version. There is also an excellent introduction dealing with the relationship between the stories and the probable motives of the Christianizers of Leif.

Mowat, Farley. *Westviking: The Ancient Norse in Greenland and North America*. Boston: Little, Brown and Co., 1965. An exercise in wringing dry the sagas in which every paragraph is translated and commented on—often to the extent of hypothesizing the conflict, anxiety, or other motivation behind the bare facts recorded. Leif's settlement is firmly located at Tickle Cove Pond in Newfoundland. Keen though it is, this book is let down by its failure to grasp the nature of the Icelandic manuscripts on which it is based.

Tornoe, Johannes K. *Norsemen Before Columbus: Early American History*. London: George Allen and Unwin, 1965. One of many analyses of the sagas, this study is particularly good on ships, sailing directions, and other details, including Norse observations of the sun and the account of the "wild grape."

T. A. Shippey

MERIWETHER LEWIS
AND
WILLIAM CLARK

Meriwether Lewis

Born: August 18, 1774; Albemarle County, Virginia
Died: October 11, 1809; Grinder's Stand, Tennessee

Lewis was coleader of the Lewis and Clark Expedition, the first party of white men to cross the North American continent from the Atlantic to the Pacific coast within the geographical limits of the present United States.

William Clark

Born: August 1, 1770; Caroline County, Virginia
Died: September 1, 1838; St. Louis, Missouri

After serving as coleader of the Lewis and Clark Expedition, Clark was for three decades one of the most important administrators of Indian affairs in the nation's history.

Early Lives

Meriwether Lewis was born August 18, 1774, on a plantation in Albmarle County, Virginia. Meriwether's father was William Lewis, who married Lucy Meriwether, for whom the explorer was named. Meriwether had an older sister and a younger brother. The first Lewises in America, who were Welsh, migrated to Virginia in the mid-seventeenth century, where the family became planters. Meriwether's father was a lieutenant during the Revolutionary War, but he drowned while on leave in 1779. Six months later, Lucy married Captain John Marks. After the war, the Marks family moved to Georgia, but Meriwether soon went back to Virginia to live with his relatives. There he attended several small schools taught by parsons and received some tutoring, but his chief interest and delight was in rambling in the woods, hunting, and observing nature. Although rather stiff and awkward as a child, Meriwether grew up to be a handsome young man.

When John Marks died in 1791, his widow returned to Virginia. She brought with her, besides Meriwether's brother and sister, a son and daughter she had borne her second husband.

A short time after his mother's return, Lewis became a soldier, as he was to remain most of his life. In 1794, he enlisted in the Virginia militia to help suppress the Whiskey Rebellion in western Pennsylvania. Liking this taste of military life, Lewis stayed in the militia until May, 1795, when he became an ensign in the United States Army. A few months thereafter, he was assigned to the "Chosen Rifle Company" which William Clark commanded, and during the short time that the two men were together, they became fast friends. Later that year, Lewis joined the First Infantry Regiment and for the next four years was engaged in a number of noncombatant duties, mainly on the Western frontier. In December, 1800, he was promoted to captain and became regimental paymaster.

It was while he was thus occupied that, in February, 1801, President-elect Thomas Jefferson wrote to invite Lewis to become his private secretary, probably with a view to naming him to command a transcontinental exploring expedition. Jefferson had thought about, and even planned for, such an undertaking since the United States had won its independence in 1783. In 1792, Lewis, then only eighteen years old, had volunteered for the assignment. Jefferson chose someone else, however, who failed to go.

Soon after coming to Washington, Lewis, under the president's direction, began to plan and prepare for the expedition. He obtained scientific and technical training from members of the faculty of the University of Pennsylvania, collected, with their advice, various kinds of equipment and supplies, and gathered information on his proposed route. Following congressional approval and funding of the mission and his formal designation as its commander, Lewis, early in 1803, with Jefferson's concurrence, invited his friend William Clark, with whom he had maintained contact since they served together in the army, to be its coleader.

Clark was born August 1, 1770, on his family's plantation in Caroline County, Virginia. He was the youngest of six sons and the ninth of ten children of John and Ann (Rogers) Clark. The Clarks had emigrated from England some time in the seventeenth century and, like the Lewises, had become planters. When the Revolution came, the Clarks were staunch patriots, and all of William's older brothers fought as officers in the War for Independence. The most famous was Brigadier General George Rogers Clark, who was the conqueror of the Illinois Country. William, who was too young to fight, stayed home. He received a little formal schooling and acquired the rudiments of learning, but mainly he developed the skills of a frontiersman: the ability to ride, hunt, and shoot.

Library of Congress

When he was fourteen years old, Clark moved with his family to a new plantation near the Falls of the Ohio at Louisville. As a young Kentucky frontiersman, Clark, a big, bluff redhead, served with the militia in several campaigns against the hostile Indian tribes living north of the Ohio River. In March, 1792, he was commissioned a lieutenant in the United States Army and two years later fought under General Anthony Wayne in the famous battle of Fallen Timbers. In July, 1796, however, Clark resigned his commission and returned home, where for the next seven years he managed his aged parents' plantation. It was there that, in July, 1803, he received Lewis' invitation to join him in leading a transcontinental exploring expedition and quickly accepted it.

Life's Work
About the time Clark received his letter, Lewis, in the East, completed his preparations for the expedition and received final detailed directions from the president. The mission's purpose, as stated by Jefferson, was to explore the Missouri River up to its source in the Rocky Mountains and descend the nearest westward-flowing stream to the Pacific in order to extend the American fur trade to the tribes inhabiting that vast area and to increase geographical knowledge of the continent. With these instructions, Lewis left Washington for Pittsburgh. Descending the Ohio River by boat, he picked up Clark at Louisville, in late summer 1803. Together with a few recruits for the expedition, the two men proceeded to Wood River, Illinois, opposite the mouth of the Missouri, where they encamped early in December. During the next five months, Lewis and Clark recruited and trained their party and finished their preparations for the journey.

With everything in readiness, the expedition set out on May 14, 1804, for the Pacific. Lewis, still a captain in the First Infantry, was the expedition's official commander. Although commissioned only a second lieutenant of artillerists, on the expedition, Clark was called "captain" and was treated in every way as Lewis' equal. During the journey, Lewis, a rather intense, moody introvert, spent much of his time alone, walking on shore, hunting, and examining the country. Because Lewis was better-trained scientifically and the more literate of the two officers, he wrote most of the scientific information recorded in the expedition's journals. Clark, a friendly, gregarious individual, spent most of his time with the men in the boats. He was the expedition's principal waterman and mapmaker, and he was better able to negotiate with the Indians. Together, the two officers' dispositions, talents, and experience complemented each other superbly. Despite the differences in their personalities, they seem always to have enjoyed the best of personal relations.

LEWIS AND CLARK EXPEDITION

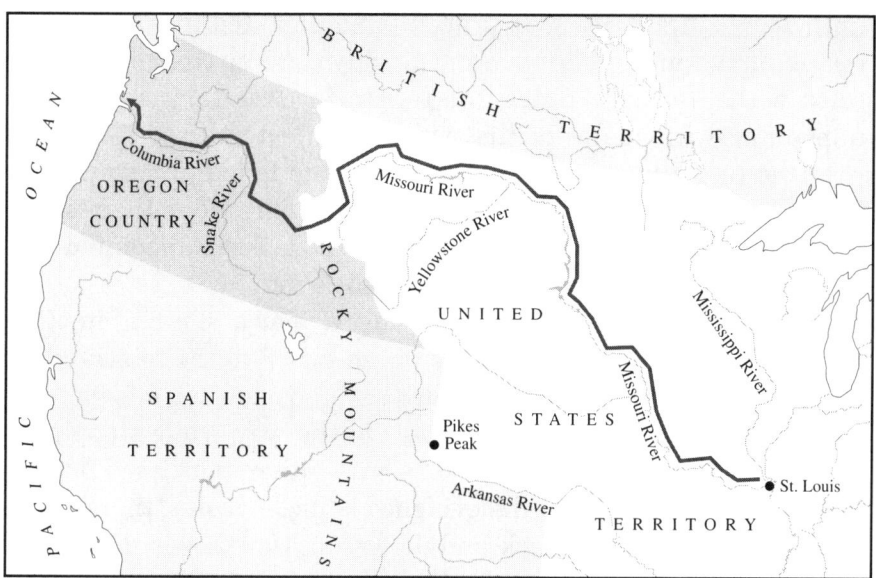

In its first season's travel, the expedition advanced some sixteen hundred miles up the Missouri and went into winter quarters in a small fort, named Mandan for the nearest Indian tribe, situated in modern North Dakota. It was here that Lewis and Clark hired Jean-Baptiste Charbonneau and, more important, his Indian wife, SACAGAWEA, as interpreters. Sacagawea also proved to be an invaluable guide and diplomat. The following spring the expedition proceeded to the headwaters of the Missouri, made a portage of the Rocky Mountains, and descended the nearest westward-flowing tributaries of the Columbia as well as the Columbia itself. Lewis and Clark reached the Pacific by mid-November, 1805. After wintering a few miles from the ocean, in a post they called Fort Clatsop, for a nearby tribe, in March, 1806, the explorers set out for home and arrived in St. Louis in September, having long since been given up for lost by virtually everyone but Jefferson.

As rewards for their great achievement, the president appointed Lewis governor of Louisiana Territory and Clark its principal Indian agent and brigadier general of the territorial militia. Detained in the East by business related to the expedition and other matters, Lewis did not actually assume the governorship of the territory until March, 1808. He soon proved to be unsuited for the office by temperament and experience and quickly ran into trouble. He quarreled with Frederick Bates, the territorial secretary, and became unpopular with many of the people of the territory. He seldom reported to his superiors in Washington and failed to consult them on his policies and plans. As a result, he fell under their severe criticism, and he probably would not have been appointed to a second term of office had he survived the first.

In September, 1809, after about only a year and a half in office, Lewis left St. Louis for Washington, in order to try to straighten out his affairs with the government and to renew his efforts to get the expedition's journals published. On the way, while stopping at a tavern on the Natchez Trace, he was either murdered or committed suicide. Although the evidence is inconclusive, there is reason to believe, as did Clark and Jefferson, that Lewis died by his own hand. Thus, at the age of thirty-five, ended the life of this great pathfinder.

Clark, in the meantime, was mainly concerned with improving relations and promoting trading activities with the Indian tribes of the territory and protecting the white settlers against the tribes of the Upper Mississippi who were allied with the British in Canada. Following Lewis' death, he was offered the governorship of Louisiana, but he declined it because he felt he lacked political experience. In June, 1813, however, the governorship of the Territory of Missouri, as the Louisiana Purchase was called after 1812, again became available, and this time Clark accepted it. During the War of 1812, which was then raging, Clark's chief responsibility was to defend the territory against the hostile Indians of the Upper Mississippi. After the war, Indian relations and the economic and political needs of the white settlers pouring into Missouri absorbed his time and interest.

Following Missouri's admission to the Union in 1821, Clark (an unsuccessful candidate to be the state's first governor), was appointed superintendent of Indian affairs at St. Louis and retained responsibility for the tribes of the Missouri and Upper Mississippi.

Clark held this office until his death on September 1, 1838. As superintendent of Indian affairs, he played a major role in effecting the removal of Indians living east of the Mississippi and in Missouri to new lands in modern eastern Kansas.

Unlike Lewis, who never married, Clark was an affectionate family man. In 1808, he married Julia Hancock, with whom he had five children. Following Julia's death, in 1821 he married her cousin Harriet Kennerly Radford, a widow, who bore him two sons. Four of his sons lived to manhood.

Impact

Lewis and Clark's fame rests almost entirely on the success of their great expedition, one of the most extensive explorations undertaken in their time. They and their companions were the first American citizens to cross the continent and the first white men to traverse it within the area of the modern United States. During a journey that lasted a little more than twenty-eight months, the expedition traveled more than eight thousand miles. On the entire trip, only one man, Sergeant Charles Floyd, lost his life, and he died from a cause almost certainly unrelated to his exploring activities. In their contacts with thousands of Indians, they had only one minor encounter, which cost the lives of two Indians. The total expense of the undertaking was a little less than forty thousand dollars. Although Lewis and Clark did not find a commercially feasible route across the continent, as Jefferson hoped they would, they did make a significant contribution to the existing knowledge of the geography of a great part of North America. They also took a historic step toward opening the Trans-Mississippi West to American trade and subsequently to American settlement, thus providing the basis for one of the United States' strongest claims to the Oregon Country. Their great achievement stimulated the pride of the American people and served to make Americans aware of the vastness of the continent on which they lived.

Although Lewis' career after the expedition was short and hardly noteworthy, Clark's was long and eminently successful. In three decades of dealing with the tribes of the Upper Mississippi and the Trans-Mississippi West, he carried out the policies of the federal government faithfully and effectively, helping to adjust relations peacefully between the Native Americans and the whites. In doing so, by the standards of his own time, he treated the Indians fairly and sympathetically and, in return, had their respect and confidence.

Bibliography

Ambrose, Stephen E. *Undaunted Courage: Meriwether Lewis, Thomas Jefferson, and the Opening of the American West*. New York: Simon and Schuster, 1996. This probing and lively biography of Lewis narrates its subject's exploration of the American West in the context of Jefferson's political and scientific ambitions. Documentation is full and easy to consult. Includes end notes, illustrations, and a bibliography.

Bakeless, John. *Lewis and Clark: Partners in Discovery*. New York: William Morrow and Co., 1947. This is a readable and relatively well-researched study, but its treatment of Lewis, and especially Clark, after the expedition is brief and sketchy.

Cavan, Seamus. *Lewis and Clark and the Route to the Pacific*. New York: Chelsea House, 1991. Part of Chelsea's juvenile World Explorers series. Includes an introduction by Michael Collins, illustrations, and a bibliography.

Cutright, Paul Russell. *Lewis and Clark: Pioneering Naturalists*. Urbana: University of Illinois Press, 1969. This volume contains a wealth of detailed information on the scientific and technical aspects of the expedition, including fauna and flora discovered, topographic features discovered or named, and Indian tribes encountered.

Dillon, Richard. *Meriwether Lewis: A Biography*. New York: Coward-McCann, 1965. The only noteworthy biography of Lewis, this somewhat sentimental and romantic work provides a relatively comprehensive treatment of the subject with emphasis on the expedition.

Jackson, Donald D., ed. *Letters of the Lewis and Clark Expedition, with Related Documents: 1783-1854*. Urbana: University of Illinois Press, 1962. This is a comprehensive collection of meticulously edited letters, memoranda, and other documents dealing with all aspects of the expedition gathered from widely scattered sources.

Lavender, David Sievert. *The Way to the Western Sea: Lewis and Clark Across the Continent.* New York: Harper & Row, 1988. Includes illustrations, maps, and a bibliography.

Lewis, Meriwether, and William Clark. *The Journals of Lewis and Clark.* Edited by Bernard De Voto. Boston: Houghton Mifflin Co., 1953. Based on the 1904-1905 eight-volume Thwaites edition of *The Original Journals of the Lewis and Clark Expedition.* This single volume provides a good, readable narrative of that great enterprise that retains its flavor.

Ronda, James P. *Lewis and Clark Among the Indians.* Lincoln: University of Nebraska Press, 1984. An important, sophisticated, and engaging ethnohistorical study, this work chronicles the daily contact between the explorers and Indians and shows that the expedition initiated important economic and diplomatic relations between them.

Steffen, James O. *William Clark: Jeffersonian Man on the Frontier.* Norman: University of Oklahoma Press, 1977. Steffen sketches selectively and very briefly Clark's life, making an occasional reference to the intellectual framework which he believes explains it.

John L. Loos

CHARLES A. LINDBERGH

Born: February 4, 1902; Detroit, Michigan
Died: August 26, 1974; Hana, Maui, Hawaii

Lindbergh's historic New York to Paris solo flight in 1927 was a turning point in aviation history, and he continued to play a major role in both civil and military aviation throughout his life.

Early Life

Charles Augustus Lindbergh was born February 4, 1902, in Detroit, Michigan, the only son of Swedish-born Charles August Lindbergh (not Augustus as sometimes incorrectly cited), a Minnesota congressman, and Evangeline Lodge Land Lindbergh, a Michigan native and chemistry teacher of English-Scotch ancestry. The elder Lindbergh, a Little Falls, Minnesota, lawyer and businessman, served as a Progressive Republican in the United States House of Representatives from 1907 to 1917, where his reform interests included such issues as banking and currency, the Midwestern farmer, and the European war. Charles August and Evangeline Lindbergh were estranged early in their marriage, but young Lindbergh regularly spent time with both parents, thus living primarily in Minnesota and Washington, D.C. The elder Lindbergh had remarried after his first wife's death and young Lindbergh had two half sisters, Lillian and Eva. In his early years, Lindbergh showed his mechanical and scientific bent when, for example, he visited the laboratory of his grandfather Charles Land (a dentist and researcher) in Michigan, and when he drove the car in his father's 1916 campaign for the United States Senate. He was graduated from Little Falls High School in 1918, and, early in the same year, began working the home farm, where he remained until the fall of 1920.

After three semesters at the University of Wisconsin, where he enrolled in the mechanical engineering program and was a member of the rifle team, Lindbergh quit school in early 1922 and became a flying student at the Nebraska Aircraft Corporation in Lincoln, Nebraska. During this period, "Slim" Lindbergh (he was six feet, three and one half inches tall) gained a reputation as an expert mechanic, parachute jumper, wing-walker, and pilot. He made several swings on the barnstorming circuit in the Midwest and Great Plains with other flying buddies, and, in 1923, he purchased his first airplane, a surplus World War I Curtiss Jenny. In 1924 and 1925, he completed United States Army Air Cadet programs at Brooks and Kelly fields in Texas and was graduated at the top of his class with the rank of second lieutenant.

Lindbergh then moved to St. Louis, Missouri, where he was head pilot for Robertson Aircraft Company and joined the Missouri National Guard unit. In April, 1926, he became one of the early pilots to carry United States mail when he began flying routes to Peoria and Chicago, Illinois, for Robertson. In order to compete for the twenty-five-thousand-dollar Orteig Prize for the first New York-to-Paris flight, Lindbergh secured financial backing from St. Louis supporters; with engineer Donald Hall, he helped to design the specially built monoplane, *The Spirit of St. Louis*, at Ryan Airlines in San Diego, California. In early May, 1927, he set a transcontinental speed record when he flew from San Diego to New York via St. Louis.

Even at this point in his life, certain characteristics about Lindbergh had emerged: a constantly inquiring mind, a total sincerity, a meticulous attention to detail and accuracy, and a sense of humor. Like his father, the reform-minded congressman and scholar, he also had a stubborn independence, a sense of courage, and a quiet personal nature.

Life's Work

Lindbergh established a milestone in aviation history, when, on May 20-21, 1927, he flew *The Spirit of St. Louis* nonstop from New York to Paris. His historic flight of 3,610 miles in thirty-three hours and thirty minutes was the first one-man crossing of the Atlantic Ocean by air. The flight was followed by an unprecedented and prolonged

Archive Photos

public response, and, overnight, Lindbergh became a world figure. After receptions in Europe, Lindbergh returned to the United States aboard the cruiser USS *Memphis*, a trip arranged by President Calvin Coolidge, and was honored in many cities. He received numerous honors and awards, including the Congressional Medal of Honor and a promotion to colonel. Lindbergh also made trips to Latin America and to Mexico, where he met Anne Spencer Morrow, the daughter of United States ambassador Dwight W. Morrow. Lindbergh and Morrow were married in 1929.

During the period of rapidly expanding aviation activity after the famous flight and through the 1930's, Lindbergh served as technical adviser to Transcontinental Air Transport (TAT, later TWA) and Pan American World Airways. In this capacity, he played a major role in the testing of new aircraft, in planning the first transcontinental route for TAT (he flew the last leg in a Ford Tri-Motor), and in developing regular transoceanic routes for Pan American. It was Lindbergh, representing TWA, for example, who demanded that the Douglas DC-1 airplane be able to take off and land safely with one engine. Ultimately, the design resulted in the legendary DC-3. The pioneer aviator was among the first to recommend the use of land planes crossing the oceans, a practice now accepted after the early use of Clipper flying boats. On international route development and mapping, Charles and Anne Lindbergh made several long test flights about the world in his Lockheed Sirius monoplane, the *Tingmissartog*, which Anne has described in her book *North to the Orient* (1935). Lindbergh also served as a consultant to the Guggenheim Fund and the United States Bureau of Aeronautics, and, when the air mail crisis occurred in 1934, he took a stand in opposition to President Franklin D. Roosevelt's decision to allow the United States Army to fly the mail.

The 1930's also brought tragedy to the Lindberghs. In 1932, their first child, Charles Augustus, Jr., was kidnapped and murdered. The extensive publicity which continued during the trial, conviction, and execution of Bruno Richard Hauptmann for the crime was so distasteful to the Lindberghs that they sought refuge in Europe in 1935. They lived in England and on an island off the coast of France, seeking privacy in rearing their family, which came to include five other children: Jon, Anne, Land, Scott, and Reeve. At this point in his life, Lindbergh's interest turned to scientific research; he worked closely with surgeon Alexis Carrel in developing a perfusion pump (frequently referred to as a mechanical heart) which was able to sustain life in animal organs outside the body, and with Robert Goddard, the father of modern rocketry, for whom Lindbergh secured important financial support.

While in Europe, Lindbergh studied European military aviation and made three major inspection trips to Germany between 1936 and 1938. After these visits, convinced of German air superiority, he warned against the growing air power of the Nazi regime. In 1939, Lindbergh returned to the United States and, at the request of General Henry Arnold, assessed United States air preparations.

With the outbreak of war in Europe, Lindbergh began his antiwar crusade. Fearing possible American involvement, he took a public stand for neutrality and later joined the isolationist America First Committee. Because of this controversy, the Lindbergh image was tarnished as political charges were made over his disagreement on American foreign policy with the Roosevelt Administration. Bolstered by his father's adamant stand against Wilsonian policies prior to American entry into World War I, Lindbergh remained firm in his views.

Following the Japanese attack on Pearl Harbor, however, Lindbergh supported his country fully when it entered the war. Only later was it known that the famous aviator, who resigned his military commission in 1941 under political pressure, had personally tested every type of fighter aircraft used by the United States in the South Pacific. Although a civilian, Lindbergh flew some fifty combat missions, passing on technical knowledge which enabled American pilots to save on fuel consumption. During the war, he was also a consultant to the Ford Motor Company at the Willow Run plant and made high-altitude chamber tests at the Mayo Clinic in Rochester, Minnesota.

After World War II, Lindbergh continued to be active in commercial and military aviation, but, increasingly, his time was devoted to two other concerns, conservation and writing. The pioneer aviator, continuing his association with Pan American (his TWA affiliation had ended in the 1930's), early advised the introduction of jets and jumbo jets, which opened a new era in air travel. During the postwar years, he served in an advisory capacity on such matters as the Berlin airlift and selection of the United States Air Force Academy site. On a long-term appointment, Lindbergh was a consultant to the Department of Defense, and he was awarded the rank of brigadier general in the Air Force Reserve in 1954 by President Dwight

D. Eisenhower. In this role, one of his most important contributions was his involvement with the structuring and implementation of the Strategic Air Command.

Devoted to the idea of world ecology and the preservation of natural resources, Lindbergh came to the conclusion that modern technology endangered the natural environment of the world—a conflict he described as civilization versus the primitive. Thus, his interests moved from science to mysticism and the study of primitive peoples. Lindbergh valued simplicity in life—the earth and sky—perhaps harking back to the roots of the Minnesota farm boy with his exposure to woods and water. Indeed, Lindbergh felt strongly about his Minnesota and Scandinavian background, and he participated in several projects concerning the Minnesota Historical Society, the biography of his father, Charles August, and the proposed Voyageur's National Park in the state.

When Lindbergh became involved with conservation, especially in his work as director with the World Wildlife Fund, he relaxed somewhat his strong aversion to the press. As early as 1948, he had warned, in his brief study *Of Flight and Life* (1948), that the human race could become a victim of its own technology. Further, according to Lindbergh, the overall quality of life should be the paramount goal of humankind. He put it simply when he wrote in 1964: "If I had to choose, I would rather have birds than airplanes." Lindbergh, aviator and technician, thus was complemented by Lindbergh, conservationist and defender of wildlife. While he encouraged the use of the Boeing 747 as an efficient aircraft, for example, he questioned the economic efficiency and environmental impact of the supersonic transport and spoke out against it during the debate in 1970. His struggle with changing values is also seen by his support of American retaliatory power during the Cold War, which was set against his worry that aviation and technology had made all people vulnerable to atomic annihilation.

Lindbergh also spent considerable time in many successful writing efforts. *We* (1927) is a brief account of the famous flight; *The Culture of Organs* (1938), written with Alexis Carrel, is a record of the research on which the two collaborated. Among his many publications in the post-World War II era are his firsthand and thorough account of the 1927 flight, *The Spirit of St. Louis* (1953), which won a Pulitzer Prize; his *Wartime Journals* (1970), drawn from extensive handwritten diaries; *Boyhood on the Upper Mississippi* (1972), an account of boyhood experiences in Minnesota; and, posthumously, *Autobiography of Values* (1977), a reflective statement on his life and concerns.

After World War II, Lindbergh lived with his family in Connecticut and then, later, in Hawaii. He continued in his duties as consultant to Pan American World Airways and to the Department of Defense and served on a number of aeronautical boards. Lindbergh died in Hana, Maui, Hawaii, on August 26, 1974.

Impact

Lindbergh is remembered first for his long and significant contributions to aviation history. From the barnstormers of the 1920's to the jumbo jets of the 1970's, Lindbergh was at the center of the immense changes that characterized aviation and aerospace technology in the twentieth century. Evidence of Lindbergh's superb technical knowledge and substantial leadership is clear, as he participated in numerous crucial decisions affecting its development. It was the 1927 flight which propelled Lindbergh to prominence, and the effects were immediate. For aviation, the historic flight launched a modern era in aviation history. More than any single event, it made the American people aware of the potential of commercial aviation, and there followed a Lindbergh "boom," with a rapid acceleration in the number of airports, pilot licenses, airlines, and airplanes in 1928-1929. While the crush of publicity was overwhelmingly favorable in 1927, Lindbergh soon came to realize that demands on his time and privacy had irreversibly changed his life. He struggled to maintain his privacy for much of the remainder of his life. From an early dislike of expressions such as Lucky Lindy and the Flying Fool, his distrust for the media deepened after the 1932 kidnapping tragedy. Yet the demanding response to Lindbergh was, in part, the history of the 1920's, an age of expanding print and broadcast journalism. Amid the sensationalism and the Prohibition experiment of the Jazz Age, Lindbergh emerged as an authentic hero to many Americans. People responded enthusiastically to the youthful

Lindbergh's individualism and modest character as well as to the new technology of the airplane.

Yet Lindbergh's influence includes more than the 1927 flight, significant as it may have been. He was not simply another flyer who set a record. Indeed, his contributions to American life in the forty-seven years between the flight and his death in 1974 included substantial activity in civil and military aviation, scientific research, and conservation. Ultimately, Lindbergh was a man both of science and of philosophical thought. His broad legacy is represented not only in aviation but also in his insistence that, if the planet is to survive, there must be an understanding between the world of science and the world of nature.

Bibliography

Bilstein, Roger E. *Flight in America, 1900-1983: From the Wrights to the Astronauts*. Baltimore: Johns Hopkins University Press, 1984. The best general scholarly treatment of American aviation. Although Lindbergh is mentioned only briefly, historian Bilstein provides the necessary framework to understand total aviation and aerospace development. Good twenty-page bibliographical note section.

Cole, Wayne S. *Charles A. Lindbergh and the Battle Against American Intervention in World War II*. New York: Harcourt Brace Jovanovich, 1974. Well-researched, scholarly study of Lindbergh's involvement in the noninterventionist movement prior to World War II. Cole, who also authored a book on the America First Committee, utilized Lindbergh interviews and the Lindbergh Papers in his work.

Crouch, Tom D., ed. *Charles A. Lindbergh: An American Life*. Washington, D.C.: Smithsonian Institution Press, 1977. Brief but informative volume based primarily on lectures delivered at the Smithsonian by John Greierson, Paul Ignatius, Richard Hallion, Wayne Cole, and Judith Schiff. It also includes notes on *The Spirit of St. Louis* by engineer Donald Hall and a reliable, selected fifteen-page bibliography.

Davis, Kenneth S. *The Hero: Charles A. Lindbergh and the American Dream*. Garden City, N.Y.: Doubleday and Co., 1959. Popular account of Lindbergh's life by a well-known journalist. As in many such accounts, there are factual inaccuracies regarding Lindbergh history, yet Davis provides a good overview and some insights into Lindbergh's life. Includes an eighty-two-page bibliographical essay.

Fisher, Jim. *The Lindbergh Case*. New Brunswick, N.J.: Rutgers University Press, 1994. This extensive work (480 pages) includes illustrations and a bibliography.

Giblin, James. *Charles A. Lindbergh: A Human Hero*. New York: Clarion Books, 1997. Although written for a juvenile audience, this is a thorough study (at 212 pages) of the controversy, tragedy, and achievement that made Lindbergh's life.

Larson, Bruce L. *Lindbergh of Minnesota: A Political Biography*. Foreword by Charles A. Lindbergh. New York: Harcourt Brace Jovanovich, 1973. Primarily a scholarly study of Lindbergh's congressman father, who had a very strong influence on the aviator's life and values. References to the younger Lindbergh and family history until the elder Lindbergh's death in 1924. Lindbergh interviews and Lindbergh Papers included in research.

Lindbergh, Anne Morrow. *Bring Me a Unicorn: Diaries and Letters of Anne Morrow Lindbergh, 1922-1928*. New York: Harcourt Brace Jovanovich, 1972.

―――――. *Hour of Gold, Hour of Lead: Diaries and Letters of Anne Morrow Lindbergh, 1929-1932*. New York: Harcourt Brace Jovanovich, 1973.

―――――. *Locked Rooms and Open Doors: Diaries and Letters of Anne Morrow Lindbergh, 1933-1935*. New York: Harcourt Brace Jovanovich, 1974.

―――――. *The Flower and the Nettle: Diaries and Letters of Anne Morrow Lindbergh, 1936-1939*. New York: Harcourt Brace Jovanovich, 1976.

―――――. *War Within and Without: Diaries and Letters of Anne Morrow Lindbergh, 1939-1944*. New York: Harcourt Brace Jovanovich, 1980. In these five volumes, Anne Morrow Lindbergh, an accomplished and recognized author, provides important documentation of and insights into the Lindbergh story in her firsthand account of the years between 1922 and 1945. Comments on specific portions of Lindbergh history may also be found in several other published works by Anne Morrow Lindbergh.

Lindbergh, Charles A. *Autobiography of Values*. Edited by William Jovanovich and Judith A. Schiff. New York: Harcourt Brace Jovanovich, 1978. San Diego: Harcourt Brace Jovanovich, 1992. Lindbergh is still the best source on Lindbergh. Published posthumously, this study was drawn from extensive manuscript material and notes written over a forty-year period. It touches on virtually all aspects of his varied life and career but strongly emphasizes Lindbergh's growing concern for the natural environment and his plea for a balance between science and nature. An essential work in understanding Lindbergh and his times. The 1992 edition includes a new introduction.

_____. *Boyhood on the Upper Mississippi: A Reminiscent Letter*. St. Paul: Minnesota Historical Society, 1972. An outgrowth of several trips to Minnesota aiding various projects on Lindbergh history, Lindbergh responded to Minnesota Historical Society director Russell W. Fridley's request for Lindbergh data with this long letter. Recounts his boyhood years.

_____. *The Spirit of St. Louis*. New York: Charles Scribner's Sons, 1953. Lindbergh's thorough account of the New York to Paris flight in 1927. This literary effort won for him the 1953 Pulitzer Prize for autobiography and biography in 1954. He writes a compelling narrative of the flight and also uses the flashback technique to touch briefly on earlier parts of his life. Lindbergh's book was the basis for the film *The Spirit of St. Louis* (1957).

_____. *The Wartime Journals of Charles A. Lindbergh*. New York: Harcourt Brace Jovanovich, 1970. Selected portions from lengthy handwritten diaries which Lindbergh kept during the wartime era between 1938 and 1945. Helpful in clarifying his involvement with the nonintervention movement, relations with the Roosevelt Administration, and his wartime activities after Pearl Harbor.

Milton, Joyce. *Loss of Eden: A Biography of Charles and Anne Morrow Lindbergh*. 4th ed. New York: HarperCollins, 1993. This lengthy (520 pages) illustrated biography includes a bibliography and an index.

Ross, Walter S. *The Last Hero: Charles A. Lindbergh*. Rev. ed. New York: Harper and Row, 1976. Popular account of Lindbergh's life written by an editor and publisher. Contains some factual inaccuracies. Book went through several editions from first publication in 1964. Broad overview for the lay reader, mostly drawn from secondary sources, with an eighteen-page note section on research.

Bruce L. Larson

DAVID LIVINGSTONE

Born: March 19, 1813; Blantyre, Scotland
Died: May 1, 1873; Chitambo's village near Lake Bangweulu,
Central Africa

Although Livingstone is often thought of primarily as a missionary, in truth he was singularly unsuccessful in this endeavor. His actual importance was as an explorer whose travels, together with moving appeals asking Britons to do something about the slave trade in the African interior, focused the eyes of the civilized world on the "dark continent."

Early Life

David Livingstone, the son of an impoverished tea vendor who was more interested in distributing religious tracts than selling tea, was born in Blantyre, near Glasgow, on March 19, 1813. Although he grew up in a very large family under adverse economic circumstances, Livingstone managed, even though he began working in a cotton mill at the age of ten, to secure a solid education. He accomplished this by studying in every spare moment, and, while still in his teens, he determined to become a medical missionary.

His preparations for such a career were successful, and when he completed his medical studies and became Dr. Livingstone in 1840, he was already unusual: It was simply unheard-of for a factory boy from a poverty-stricken background to achieve such educational heights. Livingstone's original intention had been to serve in China, but the outbreak of the Opium War effectively ended this plan. Instead, he sought and won an assignment from the nondenominational London Missionary Society to work in South Africa. He reached Cape Town on March 14, 1841, and from there made his way into the interior to the mission station of Kuruman, Bechuanaland. This isolated outpost had been established by Robert Moffat, and a

Archive Photos

few years later, in 1845, Livingstone would marry Moffat's daughter, Mary.

Life's Work
For most of the first decade that he spent in southern Africa, from 1841 to 1849, Livingstone devoted himself to the type of labors that were expected from missionaries. Livingstone founded three sepa-

rate mission stations—Mabotsa, Chonuane, and Kolobeng—in the interior to the north of Kuruman. Yet he quarreled incessantly with other missionaries and his superiors, and the sole convert he made, a chieftain named Sechele, soon lapsed. By 1849, frustrated with the routine of mission station life and increasingly attracted by the vast unexplored region to the north, he began traveling as a sort of itinerant missionary.

In truth, his religious duties were increasingly subordinated to exploration, and it was in the field of discovery that Livingstone made his real mark. Between 1849 and 1852, he made three journeys, with the noted big-game hunter William Cotton Oswell as a companion, which altered the entire course of his career. They explored Lake Ngami and the upper reaches of the Zambezi River, and during these travels, Livingstone came to recognize the extent of the internal slave trade.

He sent his wife and children, whom he regarded as impediments to his exploring ambitions, back to Great Britain in 1852, and following their departure, he continued his travels on the Zambezi. Between 1853 and 1856, he crossed the entire southern interior of Africa, traveling first to the port of Luanda, Angola, on the west coast and then later following the course of the Zambezi to Quelimane, Mozambique, on the Indian Ocean. During the latter journey, which had commenced at Linyanti, he discovered the massive falls which he named Victoria, in honor of his queen. After the completion of this journey, he returned to England for the first time since he had originally traveled to Africa.

Livingstone was already something of a geographical celebrity, thanks to contacts he had established with Sir Roderick I. Murchison, the president of the Royal Geographical Society, and articles he had written for the society's journal. It was of his great book, *Missionary Travels and Researches in South Africa* (1857), however, which brought Livingstone national fame. He completed the book shortly after returning home, and its appearance caused a sensation. He described his travels in detail and depicted the South African interior as a region which offered a fertile field for his countrymen to pursue the laudable and interconnected goals of commerce, Christianity, and civilization. So compelling was his book, together with a series of lectures that he made in England, that he secured government support for an exploratory mission on the Zambezi. Its primary purpose was to open the interior for parties of settlers who would bring British commerce and the benefits of their advanced society to the region.

Severing his formal ties with the London Missionary Society, in 1858 Livingstone returned to Africa as head of the Zambezi Expedition. This undertaking was a fiasco from start to finish. Far from being the navigable river that Livingstone had imagined it to be, the Zambezi posed all sorts of obstacles as a potential highway to the interior. The Anglican missionaries who followed him suffered greatly from the climate, and Livingstone proved woefully inadequate as a leader in the conduct of the Zambezi Expedition. He quarreled with the other members of his party, death and disease took a heavy toll on the Universities' Mission which he had encouraged to come to the region, and his wife died en route to join him. Ultimately, the British government, under growing protest, recalled him in 1863. Livingstone returned to England by way of India and arrived home on July 23, 1864.

He was no longer the conquering hero he had been in 1857, but Livingstone's ill-fated Zambezi Expedition had added appreciably to geographical knowledge of Africa. By this juncture, he had become fascinated with the controversy surrounding attempts to discover the sources of the Nile, and thenceforth he abandoned all pretense of being a missionary. He rested in England in 1865 and wrote, in collaboration with his brother Charles, a book entitled *Narrative of an Expedition to the Zambezi and Its Tributaries* (1865).

Shortly after the book was published, he made his way back to Africa, and by early 1866, Livingstone was once more in the interior. Although his announced goals were to end the slave trade and advance Christianity, he had become virtually obsessed by the watershed of the Nile. The remaining seven years of his life would be spent in an unsuccessful search for the Nile's sources, and during most of this time, he would travel without European companions.

From 1866 to 1871, Livingstone traveled in the area of Lake Nyasa and the upper reaches of the Congo (modern Zaire) River.

Among his discoveries were Lakes Mweru and Bangweulu, but these years cost him dearly in health. Indeed, when the journalist Henry Morton STANLEY found him at Ujiji on Lake Tanganyika late in 1871, Livingston was near death. With the medicines and supplies proffered by Stanley, Livingstone quickly recovered. Together they explored portions of Lake Tanganyika and ascertained that the reservoir was not a part of the Nile system.

The few weeks he spent with Livingstone influenced Stanley profoundly, and he did his best to convince the aging man to leave Africa. Livingstone refused, but he did open himself up to his traveling companion in a way that he had never done with any other person. He told Stanley of the horrible massacre of Africans by Arab slave traders that he had witnessed a few months earlier, and he shared his dreams of what he desired for the continent's future. Eventually, Stanley's anxiety to let the outside world know that he had "found" Livingstone led to his departure. The two separated at Unyanyembe (near the modern city of Tabora, Tanzania), with Livingstone determined to continue his explorations for the Nile's sources.

Livingstone's health, already seriously affected by recurrent bouts of malaria and years of unconcern for his physical state, degenerated rapidly. Tragically, Livingstone's instruments had also been damaged in transit, and in his final days, he was for all practical purposes lost. After days of being carried by the handful of "faithfuls" who continued to accompany him and being in a coma much of the time, Livingstone died in the predawn hours of May 1, 1873. Death came at a small village on the shores of Lake Bangweulu. His native companions eviscerated his body and embalmed it as best they could before beginning the long journey back to the East African coast with it. Once the party reached Zanzibar, Livingstone's remains were turned over to British authorities. Eventually, his body was taken to England and interred in Westminster Abbey on April 18, 1874.

Impact

David Livingstone was a difficult, complex individual who was always surrounded by controversy. There can be little doubt that he was an abysmal failure as a missionary, husband, and father. He lacked the patience to pursue the daily drudgery required of missionaries to Africa in his era, and he was so lacking in sympathy and understanding for his wife that he drove her into the depths of alcoholism. Similarly, at least some of his children were disillusioned by Livingstone's total lack of parental concern. Against these shortcomings stand Livingstone's compassion for Africans and his almost superhuman determination as an explorer and exponent of Great Britain's civilizing mission on the African continent.

There is no disputing Livingstone's profound influence in directing British attention to what he called the "open, running sore" of the internal slave trade in Africa. Similarly, his writings, speeches, and the very nature of this controversial man captured the public imagination. In particular, his "last journey"—the final seven years he spent in Africa—fascinated the British public. There was something exceptionally poignant about the aging, ill man, struggling against all odds and frequently incommunicado as he sought to discover the sources of the Nile.

In the aftermath of his death, Livingstone came to be viewed almost as a saint. Modern observers realize that this was a misconception. Nevertheless, he exerted a profound influence on what came to be known as "the opening up of Africa." His death inspired STANLEY to complete his unfinished geographical travels, and the attention he directed to Africa—both official and otherwise—loomed large in the "scramble" of the late 1870's and 1880's. Livingstone remains a frequently misunderstood figure who has attracted scores of biographers yet who still awaits a definitive biography. Thanks to the massive Livingstone Documentation Project launched in connection with the centenary celebrations of 1973, the wherewithal now exists at the National Library of Scotland for such a study.

Bibliography

Casada, James A. *Dr. David Livingstone and Sir Henry Morton Stanley: An Annotated Bibliography*. New York: Garland Publishing, 1977. A detailed analytical bibliography of works by and relating to Livingstone.

Clendennen, Gary W., and I. C. Cunningham, comps. *David Livingstone: A Catalogue of Documents*. Edinburgh: Livingstone Documentation Project, 1979. A full listing of all known extant Livingstone documents, with a description and their location, which is invaluable for any serious study of the man. Cunningham prepared a supplement which was published in 1985.

Holmes, Timothy. *Journey to Livingstone: Exploration of an Imperial Myth*. Edinburgh, Scotland: Canongate Press, 1993. Holmes's illustrated biography includes maps, bibliographical references, and an index.

Jeal, Tim. *Livingstone*. New York: G. P. Putnam's Sons, 1973. Although characterized by a certain overemphasis on debunking the Livingstone myth, Jeal's is perhaps the fullest, and certainly the most readable, of the many modern biographies of Livingstone.

Mackenzie, Rob. *David Livingstone: The Truth Behind the Legend*. Eastbourne, England: Kingsway Publications, 1993. At 423 pages, this sizable biography includes illustrations, maps, and a bibliography.

Ransford, Oliver. *David Livingstone: The Dark Interior*. New York: St. Martin's Press, 1978. A detailed, carefully researched biography which is marred by the author's insistence that all Livingstone's life and actions can be explained by a disease which led to wide swings in mood ranging from deep depression to great elation.

Seaver, George. *David Livingstone: His Life and Letters*. New York: Harper and Brothers, 1957. A solid life notable primarily for reprinting a number of interesting Livingstone letters.

James A. Casada

SIR ALEXANDER MACKENZIE

Born: c. 1764; Stornoway, Scotland
Died: March 12, 1820; Mulnain, near Dunkeld, Scotland

By crossing Canada in 1793, Mackenzie became the first white man north of Mexico to reach the Pacific Ocean via an overland route.

Early Life

The third of four children, Alexander Mackenzie was born about 1764 on a farm near Stornoway, Scotland, on the island of Lewis. His mother, Isabella Maciver Mackenzie, died when he was still young. When a depression struck the island of Lewis, Alexander's father, Kenneth, took him to New York in 1774. Hardly had they arrived when the early stages of the American Revolution broke out. Kenneth joined the King's Royal Regiment of New York—he was to die suddenly in 1780—and young Alexander was left in the care of aunts in New York's Mohawk Valley.

By 1778, Tories were so unpopular in the Mohawk area that Alexander was sent to school in Montreal. There he was attracted by the money and adventure afforded by the fur trade. Since Great Britain had acquired Canada in 1763, opportunities for British nationals abounded. In 1779, still a teenager, Mackenzie entered the employ of the fur-trading firm of Finlay and Gregory, which later became Gregory, MacLeod, and Company.

Life's Work

Mackenzie had worked in the Montreal office for five years when Mr. Gregory sent him to the company's trading post at Detroit. Soon Mr. MacLeod, impressed with young Mackenzie's capabilities, offered him a partnership on the condition that he go to posts in the Far West in what would become Alberta and Saskatchewan.

Although the area was rich in furs, competition among the fur-trading companies was keen and often violent. Mackenzie was in charge of the company post at Île-à-la-Crosse (Saskatchewan) from 1785 to 1787. In that latter year, Gregory, MacLeod, and Company amalgamated with the larger North West Company. Mackenzie received one share in the enlarged company, which had a total of twenty shares. He was sent to a post on the Athabasca River as second in command to Peter Pond, a trapper who had already killed at least two people. Pond was the source of much misinformation. He calculated Lake Athabasca to be seven hundred miles west of its true location. He explored the region and incorrectly believed that the large river flowing from the Great Slave Lake emptied into the Pacific Ocean. Although Pond returned to the East in 1788, Mackenzie was greatly influenced by him, stating that "the practicality of penetrating across the continent was the favorite project of my ambition." From this ambition resulted two great expeditions.

Acting on instructions from the North West Company, Mackenzie embarked on his first voyage on June 3, 1789, with four other white men and a small party of Indians. Upon leaving the Great Slave Lake, he entered the river that was to bear his name. When it became apparent that the river was flowing north to the Arctic Sea and not to the Pacific, Mackenzie decided to continue to its mouth, recording in his log that "it would satisfy people's curiosity, though not their intentions." The journey down the full length of the Mackenzie River (1,075 miles) was completed in only fourteen days. Seeing the tides and the saltwater, Mackenzie was, contrary to some reports, well aware that he had reached the Arctic Sea or some arm of it. Mackenzie's expedition started back up the river, which according to myth he called River Disappointment. The men returned safely to Fort Chipewyan on Lake Athabasca on September 12, 1789, having explored one of the greatest rivers on earth. Although his exploration was of no practical use to the North West Company, Mackenzie's efforts were appreciated; he was awarded another share in the North West Company, giving him a one-tenth interest in the business.

Even before the first expedition was completed, Mackenzie was planning a second, although four years were to pass before he could undertake it. In the interim (in 1791-1792), he went to London to

Mackenzie's Northwestern Explorations, 1789 and 1793

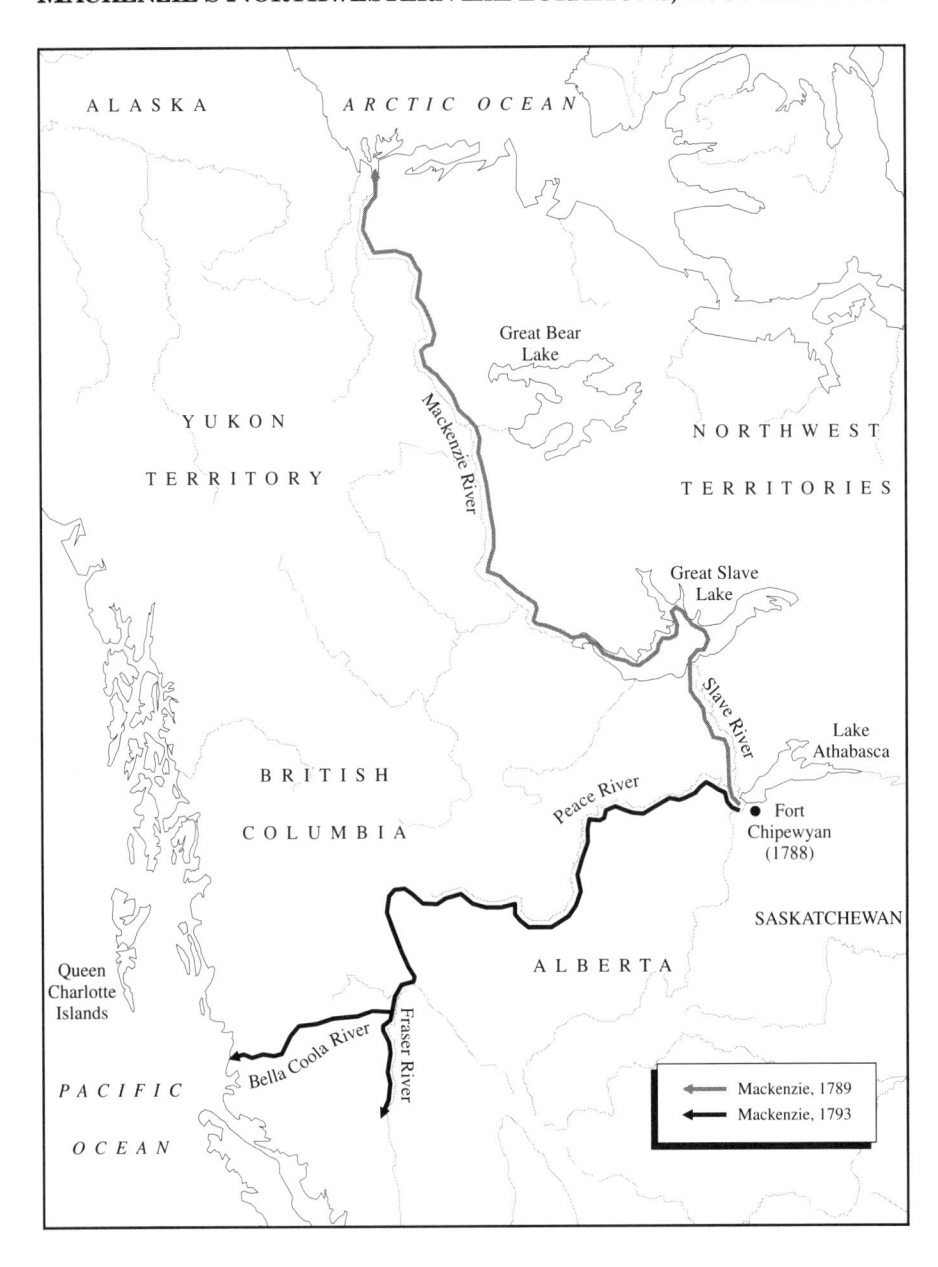

receive instructions on using astronomical apparatus, as the maps he had were useless. Despite the relative lack of equipment, even by the standards of the day, Mackenzie's observations in plotting his position on the next expedition were remarkably accurate. On October 10, 1792, he left Fort Chipewyan with the intention of spending the winter farther west and assembling a crew. At the junction of the Peace and Smoky rivers he build Fort Fork (later Peace River, West Alberta). Having assembled a modest-sized crew of nine, including two Indian interpreters, the expedition departed Fort Fork on May 9, 1793, in canoes, each twenty-five feet long and capable of carrying three thousand pounds.

Mackenzie headed due west up the Peace River. Its headwaters were shallow and exceedingly rocky, making travel difficult. Many members of the expedition urged turning back and abandoning the mission, but Mackenzie would not yield. After crossing the watershed of the Peace River over the Continental Divide to the Frazer River, he was advised by native Indians to take the shorter overland route instead of following the Frazer to its mouth. The overland route was more difficult, however, for the party was required to use Mackenzies Pass, about six thousand feet above sea level. When they descended into the Bella Coola River valley, the explorers met friendly Indians. From a high point at the Indian village, Mackenzie wrote on July 17, "I could perceive the termination of the river, and its discharge into a narrow arm of the sea." Proceeding farther, they encountered unfriendly Indians whose presence prevented any extensive exploration of the area. Nevertheless, they did canoe into the North Bentinck Arm at the mouth of the Bella Coola. There, on a large rock, Mackenzie wrote, "Alexander Mackenzie, from Canada, by land, the twenty-second of July, one thousand seven hundred and ninety-three." The following day they began the return trip to Fort Chipewyan, which they reached on August 24, having traveled more than twenty-three hundred miles.

Both expeditions are noted not only for their length but for their speed as well. Mackenzie was a man of considerable physical strength and stamina. Given the difficulties involved, he was also not easily discouraged. He expected the same from others, and they met those expectations. It is noteworthy that the wisdom of his leadership brought both expeditions back safely. At twenty-nine years of age, Mackenzie was the first white man to cross English-speaking North America—but these efforts exacted a heavy price. Mackenzie spent the winter of 1793 at Fort Chipewyan on Lake Athabasca, where he experienced a deep depression and nearly had a nervous breakdown.

The rest of his life was almost an anticlimax. Mackenzie left the West in 1794 and unsuccessfully tried to implement a plan for a unified fur trade that would include the North West Company; its longtime rival, the Hudson Bay Company; and the British East India Company. If the plan had been adopted, it would have unified the collecting and marketing of furs in the British Empire.

Mackenzie's interest in this trade plan was diverted for a time when he was offered a partnership in McTavish, Frobisher, and Company, which controlled a majority of the North West Company stock. Mackenzie and McTavish continually argued about Mackenzie's plans for the fur trade. When the partnership expired in November of 1799, Mackenzie left for England. There he wrote an account of his travels. His book, popularly known as *Voyages* (see bibliography), was published in 1801. This work attracted such immediate attention that he was knighted only two months after its publication.

Voyages also outlined Mackenzie's grand plan for the fur trade. Mackenzie even presented the plan to the British Colonial Secretary. Returning to Montreal in 1802, Mackenzie tried again to implement his trading plan through his involvement with yet another fur-trading company, but to no avail. Thwarted, Mackenzie briefly entered politics and was elected to the House of Assembly of Lower Canada, serving from 1804 to 1808. He seldom attended sessions and returned to Great Britain several times, making his permanent home there in 1810.

On April 12, 1812, at the age of forty-eight, Sir Alexander Mackenzie married the wealthy fourteen-year-old Geddes Mackenzie (no relation). At their estate at Avoch, Scotland, a daughter and two sons were born. By this time his health began to fail, and he went to Edinburgh for medical attention. On the return trip to Avoch, Mackenzie died suddenly at an inn near Dunkeld, on March 12, 1820.

Impact

In some ways, Sir Alexander Mackenzie's life could be accounted a failure. He was unable to help the Montreal-based North West Company to outflank its more centrally based rival, the Hudson's Bay Company, by finding a water route to the Pacific Ocean. When he did reach the Pacific, the route was not practical. His dream of uniting the fur trade under one cooperative venture came to nothing.

In other, more significant, ways, however, Mackenzie was a success. He explored to its full extent one of the greatest rivers on earth. He crossed North America twelve years before Lewis and Clark. Through his expeditions, and through his book, *Voyages,* Sir Alexander Mackenzie greatly enlarged humankind's knowledge of British North America. These explorations greatly strengthened Great Britain's territorial claims in the area. His efforts not only amassed a personal fortune but also gained for him recognition in the many geographical features named for him, most notably the Mackenzie River and the District of Mackenzie in Canada's Northwest Territories. They remain now as monuments to this intrepid explorer.

Bibliography

Daniells, Roy. *Alexander Mackenzie and the North West.* New York: Barnes and Noble Books, 1969. Although concentrating on his explorations, this book has a chapter treating Mackenzie's later life and a chapter assessing his achievements. Six maps; illustrated.

Gough, Barry M. *First Across the Continent: Sir Alexander Mackenzie.* Norman: University of Oklahoma Press, 1997. Volume 14 of the Oklahoma Western Biographies Series. Includes a bibliography and index.

Mackenzie, Alexander. *The Journals and Letters of Sir Alexander Mackenzie.* Edited by Kaye Lamb. Toronto: Macmillan of Canada, 1970. This book is one of a series published for the Hakluyt Society. It includes Mackenzie's diary for both expeditions and all of his letters. Contains an excellent bibliography.

_____. *Voyages from Montreal, on the River St. Laurence, Through the Continent of North America, to the Frozen and Pacific*

Oceans; in the years 1789 and 1793; with a Preliminary Account of the Rise, Progress, and Present State of the Fur Trade of That Country. Edited by William Combe. London: T. Cadell, Jr., and W. Davies, 1801. Published in New York and Philadelphia in the same year, this is Mackenzie's original work. About one-fourth of the book is devoted to a discussion of the fur trade, one-fourth to the first expedition, and one-half to the second expedition.

Sheppe, Walter. *First Man West: Alexander Mackenzie's Journal of His Voyage to the Pacific Coast of Canada in 1793*. Berkeley: University of California Press, 1962. This is Mackenzie's diary with Sheppe's intermittent explanations. Contains true maps of Mackenzie's route.

Smith, James K. *Alexander Mackenzie, Explorer: The Hero Who Failed*. New York: McGraw-Hill Ryerson, 1973. Quotes much primary source material. Appraises Mackenzie, taking into account the social and economic setting. Gives some treatment of his later life. Well illustrated.

Vail, Philip. *The Magnificent Adventures of Alexander Mackenzie*. New York: Dodd, Mead and Co., 1964. This book reads easily. It discusses Mackenzie's early life as well as his explorations but devotes only four pages to his life after 1793.

Joseph F. Rishel

FERDINAND MAGELLAN

Born: 1480; northern province of Minho, Portugal
Died: April, 1521; Mactan Island, the Philippines

Magellan was the first person to command an expedition that circumnavigated the earth. While doing so, he discovered the southernmost point of South America (later called the Strait of Magellan), was the first to sail across the Pacific Ocean (which he named), and discovered the Philippine Islands. His feat also proved that the earth is indeed round.

Early Life

Ferdinand Magellan was born in the northern Portuguese province of Minho, the third child of Dom Roy and Donha Alda Magalhães. His father was high sheriff of the district and city of Aveiro, located south of the city of Pôrto on the Atlantic coast. Magellan grew up with his siblings—sister Isabel and brother Diogo—in the Torre de Magalhães, the family farmhouse, and had a pleasant childhood in this rustic setting. At the age of seven, he attended school in the nearby monastery of Vila Nova de Mura, where he learned basic arithmetic, Latin, and the importance of harboring a strong faith in the power of Christianity.

When he was twelve, Magellan, with his father's influence, was able to travel to Lisbon and attend Queen Leonora's School of Pages with his brother, Diogo. The King of Portugal, John II, was a great supporter of marine exploration, and the young pages were expected to master such subjects as celestial navigation, cartography, and astronomy as well as the regular court subjects such as court etiquette, hunting, jousting, and swordsmanship.

In March, 1505, Magellan, his brother Diogo, and his cousin, Francisco Serrano, sailed with the fleet of Francisco de Almeida to the Orient, the three young squires having signed for a three-year service with the fleet. Magellan would serve eight years in the Orient, leaving as an extra sea hand and returning as an accomplished captain. During his service in the East, he helped establish major ports from the East African coast all the way to the Malay Peninsula. He was also involved in major confrontations with Muslim and Indian forces and was wounded several times.

Magellan, stocky in height with dark, swarthy features and piercing yet sympathetic eyes, developed strong leadership qualities and a keen appetite for adventure during his years in the East. He was also known as a fair and just man, who many times risked his life for his fellow crewmen. He was a soldier, one who could remain calm and decisive during a crisis, but one who preferred the excitement of discovery to the life of a military officer.

In July, 1511, Magellan captained a Portuguese caravel to a destination that remains unclear but was probably the Moluccas, or Spice Islands, in Indonesia, the ultimate destination of all explorers. Pepper, which was used as a food preservative by all the major countries and was therefore nearly as valuable as gold, was exported largely from the Spice Islands. During his travels, Magellan became convinced that there was an alternate route to the Spice Islands and the Indies, one that could be attained by sailing west from Europe as Christopher COLUMBUS had done. Unlike Columbus, Magellan had knowledge of a passage around the newly discovered South American continent, previously explored by a fellow navigator and friend, John of Lisbon. John had also informed Magellan that an unexplored ocean existed through the South American passage and that the Spice Islands could be reached in a few weeks time by sailing across this ocean.

When Magellan returned to Portugal in 1512, he was anxious to find backing for an expedition to discover the new sea route. He found no support from John II's successor, King Manuel I, who was much less receptive to new exploratory ventures. Rebuffed by Portugal, Magellan sailed to Spain in October, 1517, hoping to present his proposal to King Charles I. Magellan's chief contact in Spain was Diogo Barbosa, a former Portuguese navigator who had made a fortune in the spice trade and who was now the wealthy governor of the Castle of Seville. Magellan married Barbosa's daughter, Beatriz, in December, 1517. The marriage gave Magellan much pleasure, as well as a son, Rodrigo. On March 22, 1518, Magellan secured an audience with Charles. Charles was so impressed with Magellan and his proposal that he approved the expedition that same day. Preparations were then made for what would turn out to be the most epic voyage in the history of exploration.

Life's Work

During the year it took to prepare for the voyage, Magellan dealt with all details, from the rigging and loading of the ships to preventing riots and pilferage caused by spies sent from Portugal and Venice to sabotage the voyage. In the end, Magellan triumphed and finally set sail with 277 crewmen and five ships—the *Victoria*, the *Santiago*, the *Concepcion*, the *San Antonio*, and the *Trinidad*—on September 20, 1519, from San Luca, Spain, for westward passage to the Indies.

Soon after leaving Spain, Magellan's Spanish captains, led by second-in-command Juan de Cartagena, began ridiculing Magellan's authority, attempting to provoke him so that they could justify a mutiny and take command of the voyage. Magellan, however, refused to be provoked.

After suffering major storms along the African coast and disturbing doldrums near the Equator, the ships successfully crossed the Atlantic by early December, 1519. On December 8, the coast around Cape Roque in northern Brazil was sighted. Knowing that this area was under Portuguese domain, Magellan headed south into unclaimed territory and landed in what is modern Rio de

Janeiro on December 13. There the crewmen secured provisions and indulged in friendly and amorous relations with the generous natives. Two weeks later, they set sail down the coast, looking for *el paso*, the pathway first revealed to Magellan by his friend John of Lisbon.

Three months later, when no trace of *el paso* had materialized, the crew was at its breaking point. Winter storms, the worst of the expedition, began battering the ships. Magellan gave the command to seek a harbor where the ships and crew would wait out the winter for five months. The Spanish captains thought that he was mad and urged him to sail eastward for Africa's Cape of Good Hope and follow the old route to the Indies. The crew wanted him to return to the idyllic harbor at Rio de Janeiro and spend the winter there. Magellan, however, held firm.

The armada anchored at San Julián in southern Argentina on March 31, 1520. The following evening, the Spanish captains mutined. Under the leadership of Juan de Cartagena, they quickly secured three of the five ships. During a confusing boat exchange of crewmen between mutinous ships, Magellan was able to capture one of the small boatloads of men, substitute his own men for the mutiny sympathizers, and send the boat back to one of the Spanish captains along with another boat load of Magellan's men. While the Spanish captain was dealing with the first boat load, one of Magellan's men suddenly pulled out a knife and stabbed the mutinous captain as the second boat load reached the ship and scrambled on deck, ready to do battle with the rebellious crew. The crew, shocked by the sudden turn of events, became sympathetic to Magellan once more. With three ships in his favor, Magellan surrounded the other two ships and forced the remaining captains to surrender.

A trial was held for the mutineers. Two men were beheaded, and their ringleader, Juan de Cartagena, was set adrift in a small boat, never to be seen again. Through the ordeal of the mutiny, in fact through every ordeal the expedition faced, Magellan remained strong and decisive. The fact that he never doubted his ability to succeed with his mission ultimately inspired his crew to follow him, even when conditions were unbearable.

The expedition spent a total of seven months waiting for the storms to subside, first at San Julián and then farther south at Puerto Santa Cruz, where the *Santiago* was smashed against the shoreline and lost. Finally, on October 18, they set sail once again in search of *el paso*. Three days later, they came to a narrow inlet protected on either side by jagged cliffs. The inlet seemed too dangerous to navigate, but Magellan, by now appearing nearly insane to his crew, ordered the *Concepcion* and the *San Antonio* to explore the inlet. The two ships had just entered the pathway when a storm suddenly rose and swept the ships through the inlet and out of sight, while forcing Magellan's *Trinidad* and the *Victoria* out to sea. For two days, Magellan fought the storm until he was able to return to the inlet where the *Concepcion* and the *San Antonio* had disappeared. Close to panic, fearing that the two ships had been destroyed, Magellan entered the treacherous pathway. The following morning, a cloud of smoke was sighted. Then, miraculously, the two lost ships sailed into view, flags and pennants waving and crewmen cheering excitedly. They had found *el paso*.

Navigation of the strait (which Magellan called the Strait of Desire, but which was later named for him) was not complete until mid-November. During that period, the *San Antonio* disappeared. Magellan searched for the missing ship until it became apparent that the *San Antonio* had deserted and returned to Spain. Because the *San Antonio* was the largest ship and carried the bulk of their provisions, the crew urged Magellan to turn back as well. Magellan, however, would not be deterred. After the three remaining ships had sailed out of the treacherous strait and into the surprisingly calm waters of a new ocean, Magellan spoke to his men: "We are about to stand into an ocean where no ship has ever sailed before. May the ocean be always as calm and benevolent as it is today. In this hope, I name it the Mar Pacifico [peaceful sea]."

No one encountering the Pacific Ocean for the first time could have anticipated its immensity. In the following three months, nearly half of the remaining men died of starvation and scurvy. Magellan was unfortunate in that his course across the Pacific Ocean led him away from all the major groups of islands that would have supplied him with necessary provisions. During the ghastly voyage across the Pacific Ocean, Magellan threw his maps overboard in anguish, knowing that they were uselessly inaccurate. Some of the men began to believe the old superstition that the ocean would lead them not to the other side of the world but to the end of the world. When the food rotted and the water turned to scum, the dying men began eating rats and sawdust. On March 4, 1531, all the food was gone. Two days later, after ninety-eight days and thirteen thousand miles across the mightiest ocean on the planet, they reached Guam and salvation.

Taking on provisions at Guam was made difficult by the weakened state of the men and the hostility of the natives. As quickly as he could, Magellan set sail and, on March 16, found the island of Samar in the Philippines. Magellan had now achieved his personal goal, having discovered a new chain of islands for Spain. Here the expedition rested and the sick were tended, Magellan personally nursing his emaciated men.

On March 28, during the start of Easter weekend, the crew held a pageant to which the natives were invited. Magellan had made friends with the local raja and began encouraging him and his followers to convert to Christianity, which they did by the thousands. Inspired by this enthusiastic acceptance of his religion, Magellan decided not only to claim the island chain for Spain but also to convert as many natives to Christianity as he could. His desire to reach the Spice Islands, always of secondary importance to him, faded as he became more determined to make the Philippines his ultimate destination.

One month later, after exploring more of the Philippines and being favorably accepted, Magellan attempted to force the powerful raja of the island of Mactan to honor Magellan's presence. When the raja refused, Magellan assembled a small army of volunteers and the next morning, on April 27, 1521, led an attack on the raja and his followers. Because of all the hardships he had encountered and conquered, and because his expedition had now taken on a divine mission, Magellan must have come to think of himself as invincible. Unfortunately, he realized too late that he was not.

Magellan quickly realized that he and his men were hopelessly outnumbered. When he ordered a retreat, a panic ensued in which his men scrambled to the shoreline and frantically rowed back to the ships, leaving Magellan and a handful of men stranded. For more than an hour, the men defended themselves as the rest of the crew watched from the ships, until finally Magellan was struck down and killed. Antonio Pigafetta, the chronicler of the voyage, was one of the men who fought beside Magellan when he was struck down. Pigafetta was able to escape during the frenzy that followed. Later he wrote: "And so they slew our mirror, our light, our comfort and our true and only guide."

Impact

On September 8, 1522, sixteen months after Ferdinand Magellan's death, a floating wreck of a ship with an emaciated crew of eighteen men sailed into the harbor of Seville, Spain. The ship was the *Victoria*. The men, led by Juan Sebastián de Elcano, a former mutineer, staggered out of the ship and marched barefoot through the streets to the shrine of Santa Maria de la Victoria, Our Lady of Victory, the favorite shrine of their fallen leader. They lit candles and said prayers for their dead comrades, then proceeded through the streets of Seville, shocking the citizens with their wasted appearance. The *Victoria* had returned laden with riches from the Spice Islands, which had indeed been reached on November 8, 1521. As for the fate of the remaining two ships, the *Concepcion* had been burned before reaching the Spice Islands and the *Trinidad* had been captured by the Portuguese, its fifty-two crewmen hanged.

Magellan's reputation was at first defiled and degraded by his contemporaries as they learned about his behavior from the crew of the *San Antonio*, the ship that had deserted in South America. Later, however, the magnitude of his accomplishments could not be denied. He had proved that the Indies could be reached by sailing west, had discovered a pathway around the southern tip of South America, had named and crossed the largest body of ocean on the planet, had discovered a new chain of islands, had accumulated a mountain of new information about navigation, geography, and exploration, and had commanded an expedition which, after three years and forty-two thousand miles, had circumnavigated the world.

Bibliography

Cameron, Ian. *Magellan and the First Circumnavigation of the World*. New York: Saturday Review Press, 1973. Generously illustrated with maps, woodcuts, and drawings, this biography of Magellan details his life and his voyage and uses generous quotes from other biographers as well as passages from the journal of Antonio Pigafetta. Includes a selected bibliography.

Humble, Richard. *The Explorers*. Alexandria, Va.: Time-Life Books, 1978. Contains an overview of the accomplishments of the four most significant Renaissance explorers: Bartolomeu Dias, Christopher Columbus, Vasco da Gama, and Magellan. Nearly a third of the book is devoted to Magellan. Includes excellent maps and charts plus an illustrated section on the ships and navigational instruments used by the explorers, as well as a detailed description of the *Victoria*. Includes a selected bibliography.

Joyner, Tim. *Magellan*. Camden, Maine: International Marine, 1992. These 365 pages include illustrations, maps, and bibliography.

Magellan, Ferdinand, and Christopher Columbus. *To America and Around the World: The Logs of Christopher Columbus and Ferdinand Magellan*. Boston: Branden Publishing, 1990.

Parr, Charles McKew. *So Noble a Captain: The Life and Times of Ferdinand Magellan*. New York: Thomas Y. Crowell, 1953. The definitive biography of Magellan. Traces Magellan's ancestry, details the lives of all the principal men and women who affected or were affected by Magellan, vividly re-creates the time in which he lived, and chronicles his accomplishments in minute detail. Contains an extensive bibliography, including books on such related subjects as the history of Spain and Portugal, sailing-ship construction, navigation, and various locations visited by Magellan.

Pigafetta, Antonio. *Magellan's Voyage: A Narrative Account of the First Circumnavigation*. Translated by R. A. Skelton. New Haven, Conn.: Yale University Press, 1969. This is an English translation from a French text of Pigafetta's Italian journal. It is full of detailed descriptions of the events of the voyage, the lands discovered, the natives encountered and their habits and customs, and the tales told by the natives and examples of their vocabulary.

Sanderlin, George. *First Around the World: A Journal of Magellan's Voyage*. New York: Harper & Row, 1964. An interesting reconstruction of Magellan's life and voyage using letters and journals of Magellan's contemporaries. The early texts are linked by comments from the author. Most of the book is composed of excerpts from Pigafetta's journal. Illustrated, with a selected bibliography.

Stefoff, Rebecca. *Ferdinand Magellan and the Discovery of the World Ocean*. New York: Chelsea House, 1990. Written for Chelsea's World Explorers series for young readers, this work includes an introduction by Michael Collins, illustrations, and a bibliography.

Zweig, Stefan. *Conqueror of the Seas: The Story of Magellan*. Translated by Eden Paul and Cedar Paul. New York: Viking Press, 1938. A full account of Magellan's life from the time of his first voyage in 1505 to his death, and the results of his epic voyage. Contains maps and illustrations of the principal events taken from early texts.

James Kline

AL-MAS'UDI

Born: c. 890; probably Baghdad, Iraq
Died: 956; al-Fustat (Old Cairo), Egypt

A pioneer Arab historian, geographer, and chronicler, al-Mas'udi traveled extensively, gathering enormous quantities of information on poorly known lands. His work helped set the tone for future Arabic scholarship; he has been called the Herodotus of the Arabs.

Early Life

Abu al-Hasan 'Ali ibn Husain al-Mas'udi came from an Arab family in Baghdad which claimed descent from one of the early Companions of the Prophet Muhammad, though some sources erroneously describe him as of North African origin. His educational background is unknown, but his career reflects a catholic and almost insatiable thirst for knowledge.

By the standards of the tenth century, al-Mas'udi was a peerless traveler and explorer, whose feats surpass those of Marco POLO more than three centuries later. He began his travels as a young man, visiting Iran, including the cities of Kerman and Istakhr, around 915. Subsequently, he fell in with a group of merchants bound for India and Ceylon. Later, al-Mas'udi seems to have found his way as far as southern China. On his return from China, he made a reconnaissance of the East African coast as far as Madagascar, then visited Oman and other parts of southern Arabia. There followed a visit to Iran, particularly the region of the Elburz Mountains, south of the Caspian Sea.

On yet another journey, al-Mas'udi visited the Levant. He examined various ruins in Antioch and reported on relics in the possession of a Christian church in Tiberias in 943. Two years later, he returned to Syria, settling there for most of the remainder of his life. From Syria, he paid several extended visits to Egypt. Although it is uncertain whether he traveled there, al-Mas'udi's writing also demonstrates detailed knowledge of the lands of North Africa.

Al-Mas'udi's written work is characterized by his adherence to the rationalist Mutazilite school of Islamic thought. The Mutazilites, who applied logical analysis to fundamental questions of human existence and religious law, combined an intellectual disposition with a preference for vocal activism.

Life's Work

Regrettably, much of al-Mas'udi's literary work has been lost, so that in modern times it is known only by the references of others and from his own summaries in extant material. Only a single volume remains extant, for example, out of perhaps thirty that constituted al-Mas'udi's monumental attempt to write a history of the world. The surviving volume covers the myth of creation and geographical background as well as the legendary history of early Egypt.

The major work of al-Mas'udi which has survived is *Muruj al-Dhahab wa-Ma'adin al-Jawhar* (947; partial translation as *Meadows of Gold and Mines of Gems*, 1841). Apparently, there was a considerably larger, revised 956 edition of this work, but it is not extant. Al-Mas'udi laid out his philosophy of history and the natural world in *Kitab al-Tanbih w'al-Ishraf* (book of indications and revisions), a summary of his life's work.

In his books, al-Mas'udi presents a remarkable variety of information. His material on peoples and conditions on the periphery of the Islamic world is of vital importance, as modern knowledge of this aspect of Islamic history is extremely scanty. For modern scholars, however, al-Mas'udi's style and critical commentary leave something to be desired. His presentation jumps from subject to subject, without following a consistent system. Al-Mas'udi made little attempt to distinguish among his sources or to obtain original versions of information, as, for example, the eleventh century geographer/historian al-Biruni was careful to do. He treated a sailor's anecdote or a folktale in the same way as he did a map or a manuscript.

On the other hand, al-Mas'udi's uncritical approach doubtless led to the preservation of material, much of it useful, which would not have found its way into the work of a more conventional scholar. Al-Mas'udi expressed none of the condescension one sometimes finds in other writings of the time for non-Muslim authorities; he displays as much enthusiasm for learning what lay outside Islam as he does for Islamic teaching. The broad scope of his investigations was without precedent.

The juxtaposition of sources of varying authority in al-Mas'udi's work is enough to raise skeptical questions in the minds of modern readers. In discussing the geography of the Indian Ocean, for example, he first presents the "official" version, heavily dependent on erroneous ideas borrowed from Ptolemy and other Hellenistic writers, who regarded the sea as largely landlocked and accessible only through a few narrow entrances. Al-Mas'udi then lays out contrary—and more accurate—information about the Indian Ocean drawn from sailors' tales and from his own experience, indicative of the vastness of the ocean and the cultural diversity of the countries surrounding it. He also presents the orthodox notion of his time that the Caspian Sea and the Aral Sea were connected, followed by an account of his own explorations which revealed that they are separate bodies of water.

Al-Mas'udi departed from established form in presenting his information in a loosely topical manner, organized around ethnic groups, dynasties, and the reigns of important rulers instead of the year-by-year chronicle method typical of the time. In this respect, he anticipated the famed fourteenth century Islamic historian Ibn Khaldun, whose work, in turn, represents a major step toward modern historical scholarship.

A noteworthy feature of al-Mas'udi's observations of nature is his attention to geologic forces which shape the environment. Although his comments sprang mostly from intuition, they were often prescient of modern scientific theory. He wrote, for example, of physical forces changing what once was seabed into dry land and of the nature of volcanic activity.

Impact

Al-Mas'udi deserves to be included among the major Arabic historians, despite the loss of most of his work. His career marks the introduction of a new intellectual curiosity in Islam, one that sought knowledge for its own sake and paid scant attention to the boundaries between Islam and the rest of the world. His fascination with geographical elements in history and human affairs would be taken up by many later Arabic scholars.

Western historians have suggested that al-Mas'udi's intellectual disposition reflects the development of Hellenistic influence in Islamic scholarship, foreshadowing the pervasive Greek character in nontheological Islamic writing in the eleventh and twelfth centuries, particularly in Mediterranean lands. He has been compared both to Herodotus of the fifth century B.C. and to the first century A.D. Roman geographer/historian Pliny the Elder. Lack of knowledge about al-Mas'udi's training and education makes such judgments problematic, but there can be no doubt that his work is in many respects prototypical of what was to come in Islam.

Bibliography

Ahmad, S. Maqbul, and A. Rahman, eds. *Al-Mas'udi Millenary Commemoration Volume*. Aligarh, India: Indian Society for the History of Science and the Institute of Islamic Studies, Aligarh Muslim University, 1960. These essays examine the career of al-Mas'udi after one thousand years. Every major aspect of his thought and writing is covered, including his sources, his geographical and scientific ideas, his use of poetry and other devices, and his knowledge of peoples as far away as Western Europe. Several essays also discuss how al-Mas'udi's writings have been used as resources for modern scholars in various fields.

Ahmad, S. Maqbul. "Al-Mas'udi's Contribution to Medieval Arab Geography." *Islamic Culture* 27/28 (1953/1954): 61-77, 275-286. This detailed account of al-Mas'udi's life and work suggests that he was somewhat defensive about scholarship. Points out that Ptolemy was also indiscriminate about sources; al-Mas'udi may have tried to justify his eclectic sources in terms of ancient predecessors. He rejected most astronomical sources because of their reliance on astrology. Ahmed describes al-Mas'udi as a "roads and countries" scholar, heavily descriptive, less enamored with traditional cosmography.

_____. "Travels of Abu'l Hasan 'Ali B. al-Husain al-Mas'udi." *Islamic Culture* 28 (1954): 509-524. Ahmed summarizes al-Mas'udi's travels, based on his own accounts; he speculates that all travel inferences cannot be taken for granted and that some information may have been gleaned at second hand, even though the geographer's writings do not say so explicitly.

Modi, Sir Jivanji Jamshedji. "Macoudi on Volcanoes." *Journal of the Bombay Branch of the Royal Asiatic Society* 22 (1908): 135-142. Here it is shown that al-Mas'udi's descriptions and notions of volcanic activity are broadly similar to ancient ideas of hell and to myths derived from the fantastic shapes perceived in vapor clouds. Yet al-Mas'udi also displays intuition about the concentration of volcanoes in certain geographical areas and reports volcanic activity from as far away as Java and Sumatra.

Tarif, Khalidi. "Mas'udi's Lost Works: A Reconstruction of Their Content." *Journal of the American Oriental Society* 94 (1974): 35-41. Tarif discusses the growing realization among scholars of al-Mas'udi's importance. A total of thirty-four titles have been attributed to him; his historical works apparently were produced after a long period of reflection on law, philosophy, science, and theology. Thus he relied on scientific explanations for historical schema, anticipating the methods of Ibn Khaldun.

Ronald W. Davis

PEDRO MENÉNDEZ DE AVILÉS

Born: February 15, 1519; Avilés, Spain
Died: September 17, 1574; Santander, Spain

Menéndez de Avilés developed the Florida peninsula as a colony of the Spanish crown.

Early Life

One of twenty-one brothers and sisters, Pedro Menéndez de Avilés was born in the seaport town of Avilés on Spain's northern coast. A member of a family with claims to *hidalgo* (minor nobility) status, he was related also by marriage to the important Valdés clan. Like many of his relatives, friends, and contemporaries who lived in this port city, Menéndez turned to the sea in pursuit of a career.

The young seaman served initially with a leading local privateersman, Alvaro Bazán, in battles with French corsairs operating off of the coast of Western Europe. Soon he bought his own small ship with the prize money that he had earned and began the pursuit of French raiders under royal commissions granted by the Spanish crown. His successes led him to expand his operations across the Atlantic to the Indies. He became a captain-general and commander of the Spanish treasure fleets plying the routes between their colonies and the homeland.

His rising reputation as a naval leader caused Spain's Emperor Charles V to assign Menéndez to accompany the emperor's heir, young Prince Philip, to England for the latter's wedding to Mary Tudor, the eldest daughter of King Henry VIII. While the close relationship that he had developed with the heir to the throne during this period helped Menéndez throughout his subsequent career, he incurred also the enmity of the powerful merchant circle in Seville, which had to bear the expense of maintaining the armed fleet needed to protect the sea lanes to Spain's colonies. The merchants saw the young seaman as a potential rival for the profits emanating from the transatlantic trade.

Life's Work

By the middle of the sixteenth century, open hostilities had broken out between Spain and France, exacerbated by the militant Protestant movements that had swept Europe. Philip II, now king of Spain, turned south to stem the Protestant tide both on the European continent and in Spain's colonial empire.

Rumors had reached the Spanish court of French incursions and the establishment of settlements on the coast of Florida, territory claimed by Spain by right of discovery. Loss of control of this strategic area would seriously jeopardize Spanish sovereignty over the whole Caribbean zone. Philip II decided that he had to take strong countermeasures as quickly as possible. Previous attempts by Spain to establish a permanent colony under its explorers Hernando de Soto and Juan Ponce de León had ended in failure.

Philip now turned to his successful, experienced Captain General Menéndez to evaluate and to make recommendations on how to deal with the Florida problem. The veteran sailor replied quickly that the French threat was a real one and that the so-called Protestant heretics, if they were successful in enlisting Florida's Indians in their cause, could threaten the Spanish political and economic status quo throughout the area. Menéndez recommended the dispatch of an expedition immediately to rout the French if they had indeed established bases there, to institute agricultural settlements with Spanish immigrants, and to employ missionaries to convert the indigenous peoples to Catholicism. He estimated the cost of such an expedition, together with a necessary year's supplies after landing, to be in the neighborhood of eighty thousand ducats to the royal treasury.

The king and his advisers countered the proposal by offering to license Menéndez as *adelantado*, or governor, promising him lands, revenues, and titles if he would undertake the expedition largely at his own expense, but with some financial and material support by the Crown. Such an arrangement had become a common practice employed by Spanish royalty at the time, since it reduced the burden on the Crown's own finances. Menéndez accepted this risky venture involving exploration, the probabilities of serious conflict, the transfer of a substantial group of immigrants, and the religious conversion of the native population.

Although the new *adelantado* initially received moral support, military men, supplies, and missionaries to aid him, he encountered resistance from another quarter, the merchants of Seville and the Casa de Contratación—powerful groups that played a major role in the transatlantic trade between the Spain and its colonies. They delayed the aid pledged by the Crown in furnishing the ships and supplies Menéndez required to start his enterprise.

Undaunted, Menéndez not only employed all his own personal resources in the project but also secured the support of family and friends in and about Avilés. These loyal comrades became the key personnel on which he depended in building and administering the new colony. The group pledged not only their wealth but also their lives in support of their kinsman and close friend.

Meanwhile, powerful French Huguenot interests had begun assembling a fleet of their own. They did indeed have a colony started at Fort Caroline on the Florida coast, and they planned to reinforce this fledgling operation before Menéndez arrived.

By the time that the Spanish expedition had reached Florida in mid-1565, the enterprising French had already reinforced their settlement at the Caroline location. On September 8, Menéndez landed north of the French fort at a beach that he named St. Augustine and dedicated to the Spanish crown. It became the headquarters for the new *adelantamiento*, or seat of government.

Two weeks later, the Spanish leader marched south, carried out a surprise attack on Fort Caroline, captured it, and killed most of the inhabitants. Later, when some of the survivors of the initial battle who had escaped attempted to surrender, Menéndez executed the majority of them as well. Such massacres of the vanquished were all too common in the bloody encounters among European rivals. In this case, the Spaniards had quickly and forcefully ended the French threat to their control of Florida.

The *adelantado* then proceeded to launch his threefold plan of action for the colony: the establishment of military bases, the preparation for the influx of permanent settlers, and the religious conversion of the indigenous peoples. The progress proved to be slow. Food remained in short supply during the settlement's initial stages, causing low morale within the garrisons, and the local natives proved to be difficult to convert to a new religion.

The relationships between the conquistadores and their Indian charges were tumultuous. The missionaries that accompanied the soldiers insisted that the natives give up their traditional gods and adopt Catholicism exclusively. The friars demanded that the converts discontinue their practices of polygamy, sodomy, and child sacrifice, customs that were accepted traditionally within their culture. The Spanish soldiers also took by force what they wanted from the Indians and abused their women as well. Moreover, when the Spaniards adopted a particular tribe as allies, they immediately incurred the enmity of that group's traditional adversaries.

Vital supplies continued to be a problem for Menéndez's Floridian colonies. Officials both in Cádiz and in Havana either ignored the *adelantado*'s requests or demanded prepayment for goods to be delivered. The scarcity of provisions critical to the settlements' welfare kept the outposts at a bare survival level.

Despite the hazards facing the Spaniards throughout Florida, Menéndez managed to establish a string of forts along the shores of the peninsula. Unfortunately, sporadic raids by Indians, food shortages, and mutinies created problems for the Spanish leader whenever his duties called him away from the peninsula. On many occasions he was forced to punish drastically, and in some cases to execute, malefactors.

The cost involved in establishing and supplying these outposts proved to be much higher than anticipated. The colony's backers lost many ships and cargo in the process of navigating through uncharted waters. Menéndez and his associates also suffered severe financial reverses, because Florida's natural resources offered little in the way of immediate return on investment. Disorder broke out constantly among the unruly soldiery when they came to realize that there was little loot to be acquired from the native population. Menéndez decided to return to Spain and present his problems to the Crown.

Unfortunately for the Spanish colony's leader, Philip II had turned his attention to more pressing difficulties closer to home. Both France and England threatened Spain's control over its spheres of interest on the European continent itself. Accordingly, faraway Florida ranked low on the king's list of priorities.

Nevertheless, a visit to court by Menéndez did produce some favorable results. King Philip added to Menéndez's Florida command the post of governor of Cuba as well. Menéndez acquired command of a newly formed armada to operate as Spain's main line of defense throughout the Caribbean. Recalcitrant Seville and Cádiz merchants received orders from the Crown to furnish Menéndez with overdue money and supplies.

The *adelantado* was not left to govern his Florida enterprises for much longer. Philip recalled him to Europe for a new, somewhat more mysterious, assignment in mid-1573. He gave Menéndez command of a great two-hundred-ship armada, the purpose of which was to launch an attack against English home ports and to cut off supplies to English raiders harassing Spanish shipping in the Americas.

On September 17, 1574, while in the midst of organizing this undertaking, Pedro Menéndez de Avilés suddenly died, perhaps poisoned by English spies. Certainly Spain's outstanding seaman represented a serious threat to the English crown. When Philip attempted an invasion of England some fourteen years later under a less experienced and less successful admiral, the undertaking proved to be a disaster.

Although Philip II had heaped honors of all kinds on his captain general, the huge expenses of the Florida expedition left Menéndez penniless at the time of his death.

Impact

Pedro Menéndez de Avilés has been criticized by some historians for his brutal repression of the French colonists who attempted to secure Florida for their own king. He is credited with being the first European to colonize the peninsula on a permanent basis as well as to found the oldest city in the continental United States, St. Augustine. Although he sustained prohibitive financial losses personally in his attempt to develop the colony, Menéndez never wavered in his loyalty to the Spanish ruler or in his commitment to introduce Catholicism to the indigenous peoples of Florida. He must be recognized as one of Spain's outstanding colonial explorers and military leaders. He lies buried in his hometown of Avilés.

Bibliography

Barrientos, Bartolomé. *Pedro Menéndez de Avilés: Founder of Florida*. Translated by Anthony Kerrigan. Gainesville: University of Florida Press, 1965. Barrientos, a historian, was a contemporary of Menéndez.

Folmer, Henry. *The Franco-Spanish Rivalry in North America, 1524-1723*. Glendale, Calif.: Arthur H. Clark, 1954. The author attributes the Fort Caroline massacre of the French by the Spaniards to direct orders by Philip II to Menéndez to kill all of those he might find in Florida. Folmer also describes the severe reprisals that the French took against the Spaniards during their raid on the Spanish settlement that had replaced Fort Caroline in 1568.

Kenny, Michael. *The Romance of the Floridas*. 1934. Reprint. New York: AMS Press, 1970. This work is divided into two parts: "The Finding: From Ponce de León to Pedro Menéndez de Avilés, 1512-1565" and "The Founding: The Menéndez-Jesuit Period, 1565-1575." The emphasis is on the Jesuit missionary activity that took place during the Menéndez expeditions.

Lyon, Eugene. *The Enterprise of Florida: Pedro Menéndez de Avilés and the Spanish Conquest of 1565-1568*. Gainesville: University Presses of Florida, 1976. An account of the initial era of exploration and settlement of Florida by the Menéndez expeditions.

_____, ed. *Pedro Menéndez de Avilés*. New York: Garland, 1995. Volume 24 in the Spanish Borderlands Sourcebooks series, this work includes bibliographical references, illustrations, and maps.

Solís de Merás, Gonzalo. *Pedro Menéndez de Avilés*. Translated by Jeanette Thurber Connor. Gainesville: University of Florida Press, 1964. Solís de Merás was Menéndez's brother-in-law. The writer furnished an intimate knowledge of the explorer and his times.

Carl Henry Marcoux

JOHN MUIR

Born: April 21, 1838; Dunbar, Scotland
Died: December 24, 1914; Los Angeles, California

Combining his skills as a scientist, explorer, and writer, Muir played a significant role in the conservation movement and in the development of the United States National Park system.

Early Life

John Muir was born April 21, 1838, in Dunbar, Scotland. His mother, Ann Gilrye Muir, would give birth to three sons and five daughters, John being the eldest son and the third child. She married Daniel Muir, who as a child grew up under the harshest poverty imaginable. He eventually gained stature as a middle-class grain merchant and became a Presbyterian of severe Fundamentalist religious beliefs. He worshiped a God of wrath who found evil in almost every childish activity. Typically, John and his playmates would leave the yard, and his tyrannical father would fly into a rage and punish the innocent lad. When his father did not have the total devotion of his entire family, he would punish them with the greatest severity.

In 1849, at age eleven, John and his family immigrated to the United States in search of greater economic opportunity. The Muirs moved to Portage, Wisconsin, an area that had a fine reputation for wheat growing, where they purchased farmland. John marveled at the beauty of the countryside. He kept busy with farm chores and read at night when he was thought to be asleep. He also developed an early love of machinery and began the practice of waking at one in the morning to go to his cellar workshop to build things out of scraps of wood and iron. His father considered his inventions a waste of time, but John built a sawmill, weather instruments, waterwheels, and clocks. In 1860, at age twenty-two, he displayed his inventions at the state fair in Madison. His gadgets were well received, but his dour father only lectured him on the sin of vanity.

At this juncture in his life, John decided to leave home to make his own way. First, he moved to nearby Madison and attended the University of Wisconsin. He followed no particular course of study; he took classes that interested him. He seemed more concerned with learning than with earning a degree. Muir excelled in the sciences and also enjoyed the outdoor laboratory of nature. A tall, disheveled, bearded man with penetrating, glacial-blue eyes, Muir eventually grew tired of the regimentation of college. He liked books, but he loved experience more. Some men from the university were leaving to fight in the Civil War. Muir was twenty-five years old and in his junior year of school, but he decided to leave also.

From Madison, he journeyed into Canada to take odd jobs and to study the botany of the area. Later, he turned up in Indianapolis, Indiana, working in a carriage shop. With his inventive mind, he proved a success in the factory environment until one day he suffered an eye injury while working on a machine. The puncture wound affected both eyes, and soon he lost his eyesight. After a month of convalescence in a darkened room, his vision slowly returned. With a new lease on life and his eyesight fully restored, Muir decided to abandon the factory world and enjoy nature.

Life's Work

In September of 1867, Muir began a walking tour that would take him from Louisville, Kentucky, to the Gulf Coast of Florida. He found the wildlife and plants of the South fascinating. His travels took him through Kentucky, Tennessee, Georgia, and Florida, until he reached the Gulf at Cedar Key. He had no particular route planned, other than to head south. He was not disappointed in what he found on his four-month trek and decided to continue his journey. He had often read the exciting travel accounts of Alexander von Humboldt, who had explored widely in South America. That was Muir's dream also, but it was interrupted by a three-month bout with malaria. When he was almost recovered, he set off for Cuba, but, upon reaching that tropical island and after waiting for a southbound ship for a month, he settled on a new destination.

Muir believed that California offered the best climate for his malarial disorder and also afforded an environment of substantial botanical interest. He made the long journey to the West and settled in beautiful Yosemite Valley, which was snuggled in the Sierra Nevada. At times, he worked as a sheepherder and at a lumber mill, but he spent most of the time exploring the beautiful countryside, taking notes of his findings, and looking for one more glorious site of the wondrous Sierra. In 1869, Muir and a friend built a one-room cabin of pine logs near Yosemite Falls, and this became his home. He had famous visitors such as the Harvard botanist Asa Gray, the novelist Therese Yelverton, and the renowned Transcendentalist Ralph Waldo Emerson. With all, he shared the exhilarating scenes of the high country.

After four years in Yosemite Valley, Muir moved to San Francisco and dreamed of other trips. He traveled up the coast to Oregon and Washington and climbed Mount Shasta and Mount Rainier. He also made six excursions to Alaska, where he climbed mountains and studied glaciers. His favorite area was Glacier Bay in southern Alaska, but he loved any place where he could find a mountain to climb. During his stay in Alaska, he also studied the customs of the Tlingit Indians.

Muir also found time for romance. A friend introduced him to Louisa Strentzel, daughter of horticulturalist Dr. John Strentzel and owner of a large fruit ranch east of San Francisco, near the town of Martinez. Louisa and John were married on April 14, 1880. At the same time, he became the overseer of the Strentzel ranch and introduced changes that brought production to peak efficiency. Muir grafted one hundred varieties of pears and grapes onto the best strains. His effective management of the ranch provided him with economic security. For the next ten years, he neglected his writing and mountain climbing, but he and his wife grew reasonably prosperous and reared their two daughters, Wanda and Helen.

Nine years after his marriage, Muir took an important trip back to Yosemite. With him was Robert Underwood Johnson, an old friend and editor of the influential *The Century*. The two were dumbfounded by the changes that had taken place in the Sierra Nevada during such a short time. Sheep and lumberjacks had created great devastation in the valley and high country. Forest land was bare and grass root structures were severely damaged by the sharp hoofs of the sheep. Johnson was moved to action. He promised to lobby influential congressmen, and he encouraged Muir to convince the American public of their conservationist cause and the need to take action before it was too late. Muir accepted the challenge and, in two well-argued articles published in *The Century*, he convinced many readers of the desperate need to preserve some of the natural wonders of the California highlands.

In 1890, the federal government rewarded the efforts of Muir, Johnson, and other conservationists by creating Yosemite National Park. Other victories followed when Congress created Rainier, the Grand Canyon, the Petrified Forest, and parts of the Sierra as national preserves. The following year, Muir worked for the passage of legislation that eventually allowed President Benjamin Harrison to set aside thirteen million acres of forest land and President Grover Cleveland, twenty-one million acres more. Muir continued the conservationist cause by helping to create the Sierra Club in 1892. He became the club's first president, and the members vowed to preserve the natural features of the California mountains.

With the total support of his wife, Muir decided to abandon the ranch work and concentrate on furthering his writing career. In 1894, he published *The Mountains of California* and followed it with *Our National Parks* (1901), *Stickeen* (1909), *My First Summer in the Sierra* (1911), *The Yosemite* (1912), and *The Story of My Boyhood and Youth* (1913). In these works, he richly illustrated the growth of a conservationist mind and presented forceful arguments for preservation and ecological protection.

In his last years, Muir traveled to Europe, South America, and Africa, always learning and experiencing what he could. Seventy-six years of life and accomplishment came to an end in December of 1914, when Muir died in Los Angeles on Christmas Eve.

Impact

For John Muir, it had been a full life. Forced to make a decision at an early age between machines and inventions on the one hand and nature and conservation on the other, he chose the path of mountains, flowers, and preservation. In nature, he found his cathedral, and there he preached the gospel of conservation, preservation, and ecology. He walked the wilderness paths with Ralph Waldo Emerson and Theodore Roosevelt; in the end, he convinced many of his contemporaries of the rightness of his ideas.

He lived at a time when the United States was becoming a great industrial leader in the world. Still, he was able to point to the wisdom of preserving many natural wonders of the American West. While an earlier generation had plundered the East, his efforts and those of others helped to save significant portions of the West, to create large national parks and forest preserves, and to protect the ecological systems so necessary for the survival of nature.

Bibliography

Badè, William Frederic. *The Life and Letters of John Muir.* 2 vols. New York: Houghton Mifflin Co., 1924. The best collection of Muir's letters.

Cohen, Michael. *The Pathless Way: John Muir and the American Wilderness.* Madison: University of Wisconsin Press, 1984. Although there is much biographical information in this book, it is mostly an intellectual history of Muir and his ideas as he expressed them in his writings.

Fox, Stephen R. *John Muir and His Legacy: The American Conservation Movement.* Boston: Little, Brown and Co., 1981. This is a biography of Muir, a chronological history of the conservation movement from 1890 to 1975, and an analysis of what conservation means in historical terms.

Melham, Tom. *John Muir's Wild America*. Washington, D.C.: National Geographic Society, 1976. A good place to begin the study of Muir. Beautiful illustrations and sound background history.

Muir, John. *John Muir: His Life and Letters and Other Writings*. Edited by Terry Gifford. Seattle: Mountaineers, 1996. A collection that includes many of Muir's most important short writings, including studies of the Sierra Nevada and essays on California.

―――――. *To Yosemite and Beyond: Writings from the Years 1863-1875*. Edited by Robert Engberg and Donald Wesling. Salt Lake City: University of Utah Press, 1998. This selection of Muir's writings was originally published by the University of Wisconsin Press. Includes a bibliography and index.

Nash, Roderick. *Wilderness and the American Mind*. New Haven, Conn.: Yale University Press, 1967. This work traces the idea of wilderness from an early view as a moral and physical wasteland to its present acceptance as a place to preserve. John Muir emerges as one of many significant figures in this intellectual transformation.

Smith, Herbert F. *John Muir*. New York: Twayne Publishers, 1964. Approaches Muir through his writings as literary works and places him in the context of Transcendentalist literature.

Stanley, Millie. *The Heart of John Muir's World: Wisconsin, Family, and Wilderness Discovery*. Madison, Wis.: Prairie Oak Press, 1995.

Turner, Frederick. *Rediscovering America: John Muir in His Time and Ours*. New York: Viking Press, 1985. A good, sound coverage of Muir's life in the context of his times and the development of the United States.

Wilkins, Thurman. *John Muir: Apostle of Nature*. Norman: University of Oklahoma Press, 1995. Vol. 8 of The Oklahoma Western Biographies.

Wolfe, Linnie Marsh. *Son of the Wilderness: The Life of John Muir*. New York: Alfred A. Knopf, 1945. A well-written biography based on solid research that shows the many-faceted dimensions of Muir's personality.

John W. Bailey

FRIDTJOF NANSEN

Born: October 10, 1861; Store-Frøen, Norway
Died: May 13, 1930; Polhøgda, Norway

Nansen was a major Arctic explorer, an accomplished scientist, an outstanding statesman, and world-renowned for his humanitarian services to advance the rights of the oppressed and war refugees.

Early Life

Fridtjof Nansen, with his brother Alexander and a number of half brothers and sisters, grew up several miles from Christiania in a rural paradise, one with wooded areas, near lakes where they learned to swim and where, in the winter, they skated on the ice. His father, Baldur Nansen, was a lawyer of unswerving integrity, reputed to have been slender, precise, and gentle in manner, firm and honorable in character. His mother, Adelaide Wedel Jarlsberg Nansen, was a tall, industrious, and stately woman who was an accomplished snowshoer and skier, introducing her son to snowshoeing when he was four. Nansen was most like his mother, tall and large of frame, with strongly marked features and boundless energy, inheriting his mother's love of outdoor sport.

As a young man, Nansen won the national cross-country skiing championship twelve times in succession, and at eighteen he broke the world record for one-mile skating. As he grew older, it became apparent that he had acquired from his father a strong sense of obligation, a gentle manner, thoughtful sympathy for others, a careful, accurate habit of work, and a strict firmness of character. Though Nansen's family was relatively wealthy, he learned at a young age the value of hard work, discipline, and frugality. Interestingly, his very name, Fridtjof, means a Viking, or more properly speaking a "thief of peace."

As a young man Nansen became intimately knowledgeable of

Daniel Defoe's *Robinson Crusoe* (1719), Peter Christian Asbjørnsen's Norwegian fairy tales, and young fishermen bare-legged in the icy Frogner River with their bait of worms. From an early age, he never tired of boating and sailing or of boarding the sealing and whaling boats as they lay in Christiania harbor. Nansen possessed an insatiable desire for reading and asking questions—so persistently, so continually, that one friend said, "It made us absolutely ill." In later years, Nansen wrote, with a homesick longing, of the "unspeakably dear and happy home."

Life's Work

A brilliant scholar, Nansen matriculated in 1880 with honors in all natural science subjects, mathematics, and history, and in the same year entered the University of Christiania to study zoology. In the spring of 1882, while still at the university, he was asked to participate in collecting zoological specimens on an expedition to the Arctic Ocean aboard the sealer *Viking*, a six-month adventure hunting saddleback seals and one that had a momentous influence on his future career. It was during this cruise that he observed bits of driftwood and deposits of fine earth on the ice; he asked himself whether, since there were no trees in Greenland, the polar ice moved from east to west—from Siberia to Greenland—and perhaps touched the North Pole. Upon his return he was offered a position at the Bergen Museum as curator of natural history, where he spent six years.

In 1888, immediately after defending his pioneering Ph.D. dissertation in zoology, on the histology of the central nervous system of the hagfish (1887), Nansen with a party of five (including two Norwegian Lapps) made a memorable journey on skis across Greenland's inland ice sheet from east to west, which he described in his *First Crossing of Greenland* (1890). When Nansen returned home on May 30, 1889, he was greeted as a hero, having established his international reputation as a resourceful and successful Arctic explorer.

However, some explorers and scholars maintain that his major contribution to polar exploration was his expedition of 1893-1896, when he attempted to drift across the Arctic Ocean aboard the indestructible *Fram*, therefore demonstrating a water route across the polar basin. The *Fram* ("forward") was undoubtedly the most famous Arctic exploration ship, which Nansen, along with the Scottish naval architect Colin Archer, designed for its unique task of avoiding being crushed by sea ice, which was the fate of most Arctic-going vessels. This famous and unique polar vessel was a 400-ton, fore and aft ironclad, barkentine-rigged ship, one with rounded ends that allowed the ship to rise with any increased pressure of sea ice. From earlier experience in Greenland, where Nansen observed icebound driftwood, he was also aware that wreckage from the *Jeannette*, of George W. De Long's ill-fated 1879-1881 expedition which had foundered off Siberia, eventually was found along Greenland's east shore. Nansen correctly posited, from his experience on the *Viking*, that there existed a trans-Arctic ice drift. Between 1893 and 1896, Nansen's *Fram* drifted icebound from the East Siberian Sea across the Arctic basin as he had hoped, though somewhat to the left of the North Pole. He took soundings—often two miles in depth—providing a profile that dispelled the popular notion that the Polar Sea was a shallow basin.

On March 14, 1895, Nansen and his most durable skier, Lieutenant Hjalmer Johansen, left the *Fram* at 84 degrees north latitude and made a desperate attempt for the pole by dog sled and skis over the ice. In early April, at 86 degrees, 12 minutes north latitude, they gave up and turned south, arriving in Franz Josef Land, where they wintered after five months and a three-hundred-mile journey. They were rescued the following spring by a British expedition sent out to explore Franz Josef Land, led by Frederick Jackson. Two months later, on August 13, 1896, Jackson deposited Nansen and Johansen at the port of Vardø in north Norway. Unbeknown to them, the *Fram* had the same day shaken off the last of the pack ice near Spitsbergen and was steaming south for the first time in three years. Only one week after Nansen and Johansen's arrival, the *Fram* cast anchor in the far north port of Skyjervøy. Nansen had now attained the status of an oracle, for this austere, self-possessed, and enigmatic Titan proved his hypothesis about a westward drift of polar currents.

Nansen consequently exerted considerable influence upon Antarctic and Arctic explorers when he advocated the use of dogs and skis, and he recommended emulating the adaptive technology and diet of the Inuit of the Arctic. His experiences laid the basis for nearly all Arctic and even Antarctic explorations, as well as establishing him as a scholar in the relatively new fields of ethnology, nutrition, oceanography, and meteorology. These accomplishments led to his appointment as professor of zoology (1897) and professor of oceanography (1908) at the University of Christiania (now Oslo). During this period he edited a six-volume account, *The Norwegian North Polar Expedition* (published between 1900 and 1906). As a scientist, Nansen made numerous contributions, including published monographs, based on his extensive fieldwork, particularly after 1901, when he was appointed director of an international commission to study oceanographic subjects. He also made several scientific expeditions (1906-1908), mainly to the North Atlantic.

Nansen's considerable influence on the Norwegian government, in its internal affairs discussions regarding the uneasy union with Sweden, commenced his career as a statesman, for in 1905 he negotiated Norway's peaceful separation from Sweden after almost a century of Swedish rule and, before that, four centuries under Denmark. In recognition of his efforts, Nansen was appointed Norway's first ambassador to Britain (1906-1908). Nansen's international reputation was enhanced through his numerous significant contributions of informative articles to the world press and his reputation for integrity and devotion to humanitarian causes. Although Nansen wanted to continue his explorations, particularly to the South Pole, demands by his country, and later by millions of helpless World War I refugees abroad, became increasingly pressing. His strong sense of citizenship, compassion, and obligation forced Nansen to forgo his own personal ambitions in order to assist those many refugees less fortunate than himself and to perform acts of mercy without regard to his own inclinations, convenience, and even his health.

After World War I, Nansen became known internationally as a humanitarian, mainly through his services to famine-stricken Russia as well as his work in the repatriation of prisoners of war. In 1921 he was appointed as League of Nations High Commissioner for Refugees, and he was able to save millions of destitute Armenians, Greeks, and Russians, for which he received the 1922 Nobel Peace Prize, which, characteristically, he donated to international relief efforts. The League of Nations again honored him in 1931 by creating the Nansen International Office for Refugees, which won the 1938 Nobel Peace Prize. Nansen continued his work in the League of Nations, working in the Assemblies of 1925 to 1929, in which capacity he played a major role in securing the adoption of a convention against forced labor in colonial territiories, and in preparations for a disarmament conference.

Impact

Fridtjof Nansen was the most famous Arctic explorer, greatly influencing both Arctic and Antarctic explorers. In time, he was fairly judged by most explorers as one of the most successful, dedicated, innovative, and influential, if not inspirational, of all polar explorers. Much of Nansen's success in exploration was that he possessed the unique ability to plan the needs and logistics of an expedition, to choose good men, and to inspire them through his own example of dedication. Nansen's success as an explorer was based on his thorough understanding of oceanographic circulation patterns, climatology, navigation, astronomy, and Eskimo culture (especially their clothing, shelter, diet, and transportation). However, many scholars and biographers of Nansen believe that his greatest skills and contributions were demonstrated in the fields of diplomacy and politics, and of course in his tireless efforts in directing humanitarian aid to refugees.

Remarkable about Nansen are the numerous academic, literary, political, and humanitarian contributions he made. To this day, Nansen epitomizes for Norwegians a strength, integrity, and sense of both personal and national independence. His writings, in addition to those works already mentioned, included *Eskimo Life* (1958), *In Northern Mists: Arctic Exploration in Early Times* (1975), *Closing-Nets for Vertical Hands and for Vertical Towing* (1915), *Russia and Peace* (1923), and *Armenia and the Near East* (1976). In addition, Nansen wrote numerous scientific reports and journal articles, particularly in the fields of biology and zoology.

Bibliography

Bain, J. Arthur. *Life and Explorations of Fridtjof Nansen*. London: Walter Scott, 1897. An excellent biography by an Arctic explorer who knew Nansen intimately and who effectively conveys the strong sense of Norwegian nationalism evoked by Nansen's many accomplishments.

Berton, Pierre. *The Arctic Grail: The Quest for the North West Passage and the North Pole, 1818-1909*. New York: Penguin, 1988. A thorough and exceptionally well written account that critically compares Nansen's philosophy and techniques of exploration with those of Robert McClure, Charles Francis Hall, William Parry, Robert Peary, and Roald Amundsen.

Cherry-Garrard, Apsley. *The Worst Journey in the World*. New York: Carroll & Graf, 1989. Perhaps the best-written account of the Scott Expedition to the South Pole, also containing invaluable information regarding Nansen's solution for scurvy and his innovation of arctic equipment.

Christopersen, A. R. *Fridtjof Nansen: A Life in the Service of Science and Humanity*. Oslo: Cultural Office of the Norwegian Ministry of Foreign Affairs, 1961. A brief but succinct monograph that presents an in-depth and relatively detailed essay on Nansen's successful efforts in the areas of politics and diplomacy, particularly his role in the repatriation of prisoners of war and his contributions to assisting refuges of the Russian famine.

Hall, Anna Gertrude. *Nansen*. New York: Viking Press, 1940. The book presents important biographical information from Nansen's unpublished diaries, mostly on his youth and period of exploration, particularly his time aboard the *Viking* and the *Fram*.

Huntford, Roland. *Scott and Amundson: The Race to the South Pole*. New York: G. P. Putnam's Sons, 1980. A brilliant and well-researched account of previously unknown material dealing with Nansen's professional and personal relation with two famous explorers, Roald Amundsen and Robert Scott.

Mirsky, Jeannette. *To the Arctic! The Story of Northern Exploration from Earliest Times to the Present*. New York: Alfred A. Knopf, 1948. A complete survey that provides a thorough understanding of the ineradicable effect that Nansen had on both Arctic and Antarctic exploration.

Nansen, Fridtjof. *Adventure and Other Papers*. North Stratford, Conn.: AMS Press, 1977. A collection of papers that dwell mostly on Nansen's exploration, with excellent accounts of his preparations, strategies, and the logistics of his explorations.

———. *Farthest North: Being the Record of a Voyage of Exploration of the Ship "Fram" 1893-96 and of a Fifteen Months' Sleigh Journey by Dr. Nansen and Lieut. Johansen*. 2 vols. London: Harper & Brothers, 1897. An honest and even humble account of Nansen's early childhood, personal training, and devotion to his many pursuits, which were critical to his political ambitions and success as an explorer.

Ryne, Linn. *Fridtjof Nansen*. Oslo: Norwegian Ministry of Foreign Affairs, 1996. A brief but valuable account of information usually available only in untranslated books and articles.

John Alan Ross

JAMES EDWARD OGLETHORPE

Born: December 22, 1696; London, England
Died: June 30, 1785; Cranham Hall, Essex, England

With his social vision, promotional genius, military ability, and personal guidance, Oglethorpe established the colony of Georgia and frustrated the Spanish effort to push the British out of southeastern North America.

Early Life

James Edward Oglethorpe was born in London, the seventh and last child of Sir Theophilus and Lady Eleanor Wall Oglethorpe, two Jacobites who surrounded young Oglethorpe with intrigue and endowed him with the family's strong moral courage, conviction, loyalty to the Crown, and military and parliamentary tradition. Oglethorpe received the education of an English gentleman, first at Eton and then at Corpus Christi College, Oxford, where Jacobite sentiment was strong. Oglethorpe then held a commission in the British army but resigned to join Prince Eugene of Savoy in fighting the Turks. He gained a reputation for military prowess at the Battle of Belgrade (1717). After a brief Jacobite flirtation at Saint

Germain, France, where his widowed mother and sisters attended the pretender James III (also known in history as James the Old Pretender), Oglethorpe returned to the family estate of Westbrook at Godalming in Surrey. The move ended his Jacobite interest.

In 1722, Oglethorpe was elected to Parliament, succeeding his father and two elder brothers as representative for Haslemere, a seat he would hold for thirty-two years. In Parliament, Oglethorpe shook off suspicions about his Jacobitism. He won respect for integrity and hard work, and, more important, for his ambitions and interests, he cultivated several powerful friends. In Parliament, Oglethorpe opposed royal extravagance and the machinations of Robert Walpole and advocated naval preparedness, mercantile and colonial expansion, relief for the oppressed, and, later, the Industrial Revolution. Oglethorpe's humanitarianism, probably a product of his family's high-mindedness, first appeared in *The Sailor's Advocate* (1728), an anonymously published pamphlet attacking the Royal Navy's practice of impressment. The pamphlet went through eight editions. Throughout his life, Oglethorpe also professed antislavery beliefs. It was Oglethorpe's interest in penal conditions, however, that led him to his life's work.

Life's Work

In 1729, Oglethorpe was named chairman of a committee to inquire into the state of England's jails. In three reports issued in 1730, the committee cataloged the abuses of debtors' prisons. The reports electrified the public, in part because of their lurid detail and in part because such exposés were rare in an indifferent age. Oglethorpe's investigation convinced him that the nation and the debtors would be better served by settling the debtors in British colonies. There they could render service to the Crown by colonizing and defending new territory and producing crops and other goods needed in the mother country, while remaking their own lives. Such an argument was hardly new to England in the eighteenth century, for since the Elizabethan age colonizers had promised similar benefits. What gave Oglethorpe's appeal energy was the renewed public interest generated by his reports on penal conditions and his friendship with such influential men at court and in Parliament as John Lord

Viscount Perceval (later the first Earl of Egmont) and Dr. Thomas Bray (founder of several religious and philanthropic societies), who shared his interest in reform and in America.

Oglethorpe, Egmont, and eighteen other associates received a charter in June, 1732, creating the "Trustees for establishing the colony of Georgia in America." The proprietary grant was for a period of twenty-one years, after which the colony would revert to the Crown. The associates benefited from the British government's interest in placing a buffer colony on Carolina's southern frontier to protect against French, Spanish, and Indian attacks and also from its desire to increase imperial trade and navigation. Relief for domestic unemployment was a third consideration, but it lagged behind the former two. Indeed, the interest of defense and the production of exotic crops and naval stores for the mother country so outweighed the humanitarian objective that few debtors were actually recruited for the colony.

Oglethorpe quickly proved himself an energetic promoter for a project that would evoke the most vigorous and extravagant promotional literature in the British North American experience. In 1732, at his own expense, he published *A New and Accurate Account of the Provinces of South Carolina and Georgia*, stressing the commercial and agricultural advantages of the colony. Georgia's strategic position, combined with Oglethorpe's and the Trustees' appeals, helped secure regular financial support from Parliament. When his mother died in 1732, leaving Oglethorpe free of domestic responsibilities, he decided to accompany the first group of settlers to the colony—a move that fundamentally influenced the colony's development.

In November, 1732, Oglethorpe and 116 emigrants set sail for Georgia on the *Anne*. Arriving in America in January, 1733, after a successful voyage, Oglethorpe directed the settlers to the Savannah River. There he chose the site for the principal city. Oglethorpe conciliated the local Indians, securing from them both a grant for the land and an agreement whereby they would cut their ties to the French and Spanish. He laid out Savannah's distinctive pattern of squares and grids, which dominates the city even today, and then parceled out the land according to the Trustees' system of entailed grants designed to hold the settler to the soil. The cumbersome land system—which prohibited the holder from selling his property or bequeathing it to any but a male heir—would cause much trouble soon enough, but Oglethorpe imposed military discipline on the first settlers. He made a treaty with the Lower Creeks and fortified the southern reaches of the colony.

In 1734, Oglethorpe set out for England to answer charges that he was overspending and being uncommunicative. Accompanied by several Indians, Oglethorpe received an ecstatic public welcome. The press revived interest in the colony. Strengthened by the public showing, Oglethorpe gained additional support from the Trustees, including new restrictions on the colony that prohibited the sale of rum and black slavery and regulated the Indian trade through a licensing system. Meanwhile, Oglethorpe's policy of religious toleration encouraged other emigrants to join the experiment—a policy that, in 1734, led a group of Salzburger Lutherans to seek asylum in Georgia. Other German groups followed, including subsequent contingents of Salzburgers and Swiss Moravians, and Scotch Highlander Presbyterians. The British government was cool toward Oglethorpe's efforts to attract non-British emigrants, but Oglethorpe persisted. The colony needed people.

It also needed Oglethorpe's attention. Rumors of insurrection drew Oglethorpe back to Georgia in 1735. He brought John and Charles Wesley and a new batch of settlers with him. The Wesleys soon fell into disputes with Oglethorpe and the settlers and returned to England. Oglethorpe did better with George Whitefield, who came later and established an orphanage that Oglethorpe supported. Oglethorpe found the Georgia government in disarray. Lines of authority were blurred and the Trustees retained essential power in their hands, but Oglethorpe's unwillingness to delegate authority hardly helped matters. Oglethorpe further fanned the colony's troubles by his own intransigence. Vain and unbending, he insisted on enforcing the new restrictions he brought from London and honoring the Trustees' unpopular land policy. Traders from South Carolina resented the licensing system for the Indian trade, farmers chafed at restrictions on establishing a plantation-style agriculture, and the Spanish complained about Oglethorpe's southward movement, which included a new settlement at Frederica in

1736 and a fort on Cumberland Island soon after. In London, malcontents from Georgia told tales of incompetence and venality in Oglethorpe's administration. Oglethorpe responded by going to London, where he pacified the Trustees and answered all charges. He returned to Georgia in 1738 with a regiment of soldiers that he had raised at his own expense.

Military matters thereafter preoccupied Oglethorpe in Georgia. With war between Spain and England imminent, Oglethorpe repaired relations with the Indians. He persuaded the Chickasaw and Lower Creeks not to join Britain's enemies should war occur and even settled differences between the Creeks and Choctaw. He also put down a mutiny among his own men, personally grabbing the ringleaders as they shot at him. When Parliament declared war on Spain in 1739, a war, known as the War of Jenkins' Ear, that eventually became part of the larger War of Austrian Succession, Oglethorpe moved rapidly. He led a futile attack on St. Augustine in 1740, which failed partly from Oglethorpe's indecision. Although personally brave, Oglethorpe had little experience commanding a military expedition. His inability to distinguish between the trivial and the significant—a trait that afflicted his civil administration as well—further embarrassed his campaign. Oglethorpe redeemed his military reputation in 1742 when, in a series of skirmishes known collectively as the Battle of Bloody Marsh, he and his men rebuffed a superior Spanish force invading St. Simons Island. The Spanish withdrew their army from Georgia, never again to threaten seriously the British presence in North America. In 1743, Oglethorpe made another unsuccessful feint against St. Augustine, but by then Georgia was safe and Oglethorpe's American career was ending.

Civil discontent in the colony had distracted Oglethorpe while he fought to save the empire. Colonists ignored the Trustees' regulations, malcontents launched new campaigns against Oglethorpe and the Trustees in England, and the Moravians left for Pennsylvania rather than bear arms in Georgia's defense. Questions of finance especially nagged Oglethorpe. His own expenses became entangled with those of the colony, for he had borrowed against his English property to pay for Georgia's defense—money for which he would be only partially reimbursed. To add to Oglethorpe's problems, the colony storekeeper had kept poor accounts and made unwarranted expenditures. In 1740, the Trustees limited Oglethorpe's civil responsibilities so that he could concentrate on military matters.

In 1743, Oglethorpe went to England to respond to criticism and to answer charges brought by a subordinate to a court-martial. He was exonerated, but his colonizing days were over. He never returned to America. Georgia was going its own way already. Indicative of Oglethorpe's declining influence in Georgia's future was the Trustees' decision in 1750 to remove the restrictions on rum and slavery and to accept Georgia's development along the lines of South Carolina as a slave-based plantation society.

Oglethorpe married Elizabeth Wright, heiress of Cranham Hall, Essex, in 1744. The match gave him a fortune and the country estate where he lived for the rest of his life. He fought against the Jacobites in 1745, but rumors of his family's Jacobite associations trailed after him and led to charges of misconduct in not pursuing the retreating Jacobites vigorously enough at Lancashire. Oglethorpe was acquitted, but his military career was over. Using an assumed name, however, he did fight on the Continent against the French during the Seven Years' War, and did earn the friendship of William Pitt for his endeavors and promotions to general in 1765. In Parliament, Oglethorpe became something of a liberal Whig free-lance, distrusting the Hanoverian ministers, supporting civil rights for religious dissenters in the Colonies, attacking arbitrary power, and associating with the antislavery movement in England. After he lost his seat in 1754, Oglethorpe retired from public life. Oglethorpe devoted attention to his estate and to literary and artistic circles, where he became friends with Samuel Johnson, James Boswell, David Garrick, Sir Joshua Reynolds, Hannah More, and Edmund Burke, among others. He died in 1785.

Impact

From his earliest colonizing promotionals, Oglethorpe had recognized the place of Georgia in the larger British North American schema. Indeed, imperial considerations of defense and commerce, more than humanitarianism, made Georgia possible. Oglethorpe's negotiations with powerful Indian tribes marked the growing recognition among British administrators and settlers that European rivalries in southeastern North America dictated accommodations with the Native American population which held the balance of power. However clumsy, Oglethorpe's military moves underscored the fact that in the eighteenth century England would have to fight for territory on North America. Parliament's willingness to underwrite Georgia bespoke the growing strategic importance of the North American colonies in Great Britain's imperial design. In the age of imperial rivalries, visionaries needed also to be soldiers. Oglethorpe's repulsion of the Spanish at Bloody Marsh in 1742 effectively ended Spanish incursions in the southern mainland colonies and secured Great Britain's southern frontier.

Oglethorpe was the last of the great proprietary colonizers in British North America. Like William PENN, he was a visionary imbued with a strong sense of mission. Oglethorpe's promotion of Georgia captured anew the prospect of America's destiny, and like Penn, it included recruitment of non-British settlers to promise a New World elysium out of religious and cultural diversity. Unlike Penn, Oglethorpe did not temper his social vision sufficiently with practicality. Although a gentle and even generous man, Oglethorpe bridled at criticism and was egotistic and self-righteous. He never fully adapted to the democratizing tendencies of colonial life, preferring to impose rules on his charges rather than take them into his confidence. Where Penn acceded to local demands for greater self-governance, Oglethorpe insisted on compliance with all regulations. A country whig in temperament and politics in England, Oglethorpe unwittingly played the autocrat in America. His life in Georgia demonstrated how much the British colonial establishment in the late seventeenth through the early eighteenth centuries rested on the energy and enthusiasm of powerful individuals. It also served to show the limits of Old World authority in the New. Oglethorpe had founded Georgia, protected it, and given it purpose, but he could not control the social, economic, and political impulses of diverse peoples in a setting that demanded popular participation and promised individual wealth. To have done so would have defeated the idea of America that inspired Oglethorpe to believe in the Georgia experiment, and the people to risk it.

Removed from the hurly-burly of Georgia, Oglethorpe seems to have understood that fact himself. He championed America's sons of liberty during the American Revolution, and before his death in 1785 he called on John Adams, the United States minister to England, acknowledging America's promise as England's, indeed Europe's, own redemption. By his continued hope for America, the old soldier did not die.

Bibliography

Boorstin, Daniel J. *The Americans: The Colonial Experience.* New York: Random House, 1958. Boorstin's influential treatment of Oglethorpe and Colonial Georgia criticizes the Trustees for their inability to adapt to the American environment. Boorstin finds in the failure of Oglethorpe's and the Trustees' vision of Georgia the clue to the success of other forms of community in America. By comparing the Georgia experiment with those of Massachusetts and Pennsylvania, Boorstin places Oglethorpe's thought and actions in the context of American utopianism.

Church, Leslie Frederic. *Oglethorpe: A Study of Philanthropy in England and Georgia.* London: Epworth Press, 1932. This older study remains valuable for its detail on Oglethorpe's philanthropic interests, his ties to religious figures and interest, especially the Wesleys, and his social and political connections in England.

Coleman, Kenneth. *Colonial Georgia: A History.* New York: Charles Scribner's Sons, 1976. A valuable synthesis of Georgia history, Coleman's account offers an excellent brief introduction to Oglethorpe's ideas and actions and how a multireligious colony developed from his policies. Coleman is especially good at relating the politics of Colonial Georgia and Oglethorpe's and the Trustees' ineffective governance.

Ettinger, Amos Aschbach. *James Edward Oglethorpe: Imperial Idealist.* Oxford: The Clarendon Press, 1936. Ettinger's lively and sympathetic account is the fullest and best biography of Oglethorpe. Ettinger approached Oglethorpe in the tradition of George Macauley Trevelyan, who believed the eighteenth century was the age of the individual. As such, Ettinger found Oglethorpe's personality and interest formed from his family traditions of loyalty to the Crown, military service, and parliamentary responsibility. Although the bulk of Ettinger's biography focuses on Oglethorpe's American experience, the book also details Oglethorpe's Jacobite connections, parliamentary career, and literary friendships. Still, Ettinger's narrative views Oglethorpe's post-Georgia years as an anticlimax for the "imperial idealist."

Lane, Mills, ed. *General Oglethorpe's Georgia*. Savannah: Beehive Press, 1975. Lane collects and publishes a very good sampling of Oglethorpe's letters relating to the Georgia years and provides a useful introduction to Oglethorpe and Georgia, including accounts of Colonial discontent.

Spalding, Phinizy. *Oglethorpe in America*. Chicago: University of Chicago Press, 1977. Spalding reassesses Oglethorpe's life in the light of the many new materials available since Ettinger completed his research. In a balanced account of Oglethorpe, Spalding weighs Oglethorpe's ideas against his actions. He argues that Oglethorpe was not blind to American realities and that his ideas regarding a yeoman society were not necessarily doomed by the American environment.

Spalding, Phinizy, and Harvey H. Jackson, eds. *Oglethorpe in Perspective: Georgia's Founder After Two Hundred Years*. Tuscaloosa: University of Alabama Press, 1989. A scholarly attempt to place Oglethorpe in his times with the advantage of historical hindsight.

Ver Steeg, Clarence L. *Origins of a Southern Mosaic: Studies of Early Carolina and Georgia*. Athens: University of Georgia Press, 1975. In his important and provocative examination of the origins of Georgia, Ver Steeg discards most previous interpretations and argues that, although strategic considerations loomed largest in shaping policies toward Georgia, each Trustee had his own motives regarding the colony's settlement and development. In the absence of any grand design, Oglethorpe had to contend with the contradictions among both Trustees and settlers about Georgia's purpose.

Randall M. Miller

MUNGO PARK

Born: September 10, 1771; near Selkirk, Scotland
Died: 1806; near Bussa on the Niger River

Combining great ambition with tremendous courage and stamina, Park explored, and died in his efforts to traverse, the Niger River in Western Africa.

Early Life

Mungo Park was born on September 10, 1771, at Foulshiels Farm on the estate of the Duke of Buccleuch near Selkirk, Scotland. He was the seventh child of a well-to-do farmer, also called Mungo. Park received his early education at home and in the Selkirk grammar school. In 1786, he was placed as an apprentice to the Selkirk surgeon Dr. Thomas Anderson. This was a disappointment to his father, who wanted him to enter the ministry. With the help of Dr. Anderson, Park entered the medical school at Edinburgh University. He passed three sessions of medical studies and earned distinction in botanical studies. In 1791, after completing his medical studies, Park moved to London to seek employment.

Park's brother-in-law, James Dickson, a London botanist, introduced him to Sir Joseph BANKS, President of the Royal Society, who secured for him an appointment as assistant medical officer on the East India Company ship, the *Worcester*. He sailed to the island of Sumatra in February, 1792, where he collected rare plants. Park's relationship with Banks continued to develop when he returned in 1793 with his specimens and data. After presenting several papers, Park, acting on the advice of Banks, offered his services to the African Association, an organization formed in 1788 to further geographical studies of Africa.

Banks was the most influential member of the Association, and he favored Park as the successor to Major Daniel Houghton, who had disappeared on the Association expedition in 1790 to locate the course of the Niger River. The Association was impressed by Park's medical, botanical, and geographic skills as well as his physical condition for such a demanding journey. Tall and handsome in a

Archive Photos

well-chiseled way, Park possessed remarkable stamina that permitted him to perform feats of physical endurance and survive illnesses that would prove fatal to lesser men. Women found him very attractive, which proved to be important because their kindness helped him several times on his expeditions. Park's reserved personality, religious fatalism, and driving desire for eminence made him the perfect explorer, capable of pursuing success with a single-minded ambition and a certain cold-bloodedness. Park's instructions from the Association were to explore the Niger River and to gather information about the nations that inhabited its banks. He received fifteen shillings for each day he spent in Africa and two hundred pounds for expenses.

Life's Work

Park sailed from Portsmouth on May 22, 1795, aboard the *Endeavor*, a brig bound for the Gambia River for ivory. He arrived at the British factory of Pisania on the Gambia on July 5 and resided at the home of Dr. John Laidley for five months while he studied the Mandingo language and recovered from his first bout with fever. Unable to travel with a caravan, Park set out on December 2 with an English-speaking Mandingo former slave, a young servant, and his equipment. He followed Houghton's earlier route and was forced to trade off most of his trafficable goods to gain the friendship of the petty chiefs.

Danger arose when Park entered the Islamic African kingdoms. He reached Jarra in the Moorish kingdom of Ludamar before Christmas and discovered that it was the village where Major Houghton had been murdered. As he crossed Ludamar, Park was constantly abused by the people he encountered, until he was seized by Moors and taken to the residence of King Ali of Ludamar. He was held prisoner for three months while suffering humiliating treatment from his captors. In July, 1796, Park escaped through the assistance of some native women who befriended him. With only his pocket compass and a horse, he endured incredible hardships before reaching Ségou on the Niger River on July 20. He described the Niger as being as broad as the Thames River at Westminster. From Ségou, he journeyed downriver to Silla, thus proving that the Niger flowed eastward; he was forced to turn back, though, because he could no longer obtain food.

Park started back from Silla on August 3 by another route farther south, where he was again ostracized or mistreated by the natives before, nearly dead, he reached Kamalia on foot on September 16. He spent seven months during the rainy season with a native slave-trader who took him on to Pisania in June, 1797. Park sailed from the Gambia on June 15 as ship surgeon on the *Charleston*, an American slave ship bound for the Carolinas. Switching ships at Antigua, Park arrived at Falmouth, England, on December 22.

Unannounced, Park arrived in London on Christmas morning and was warmly welcomed by BANKS and the Africa Association. He had been gone for more than three years and was believed dead. His return was sensational in itself, but the news of his journey to the Niger created national excitement. Supported by a salary extension from the Association, Park wrote *Travels in the Interior of Africa* (1799), which rapidly sold out through several editions. The book was written in a dramatic and excellent literary style which made Park's name a household word and produced royalties in excess of one thousand pounds. He returned home to the Scottish countryside and soon married Alice Anderson, daughter of his old master, Dr. Anderson of Selkirk. After living at Foulshiels for nearly two years, Park established a medical practice in the village of Peebles in 1801. He refused an offer from Banks to lead an expedition to Australia because the salary was too small. In the end, however, Park's restlessness at Peebles, as well as Banks's persistence, led him to consider a new offer to return to Africa to lead an officially sponsored government expedition.

This new expedition was originally part of a larger plan by the British government to expel the French from the region of Senegambia and to establish a permanent British presence in that area. There were to be three wings to the operation—commercial, military, and naval—for the purpose of destroying French factories in Senegambia and replacing them with British factories at Wulli and Bondu. Park, as leader of the commercial wing, was to establish the new factories and negotiate trade agreements with the tribes he encountered during his exploration of the Niger. This plan was drastically altered by a change in the British government in 1804. Lord Hobart, who had approved the original plan, was replaced by Lord Camden as colonial secretary. The expeditionary force, including Park's command, was whittled down by Lord Camden.

When Park left for Africa aboard the *Crescent* on January 30, 1805, he held the rank and pay of captain and the privileges of a British envoy. He was to make treaties establishing British trading stations along the Niger while trying to discover the course of the Niger and ascertain if it were navigable from the sea. Park was accompanied by his brother-in-law, Alexander Anderson, as second-in-command, and George Scott, a Selkirk friend, as draftsman. In addition, five thousand pounds was placed at his disposal by the Treasury, and his wife and four children were guaranteed one hundred pounds a year if he failed to return.

Park's entourage, which included four carpenters and two sailors, arrived at Goree on March 28, 1805, where they were joined by Lieutenant John Martyn and thirty-five volunteers from the Royal Africa Corps. The carpenters were to build a forty-foot boat for the expedition when they reached the Niger. This expedition seemed to be efficiently organized, but it had been Park's single-minded determination and endurance that made his previous expedition a success. The size of his second expedition would become a hindrance that could not maintain the grueling pace that Park set.

So began Park's second and fatal expedition. He became impatient and against all advice led his columns into the West Africa bush during the rainy season. Sailing up the Gambia, Park reached Kayee, where he engaged a Mandingo guide named Isaaco. The overland march to Pisania taught Park that an expedition produced many different problems from traveling alone. The first rain fell on June 10, and the soldiers began to contract fevers. When possible, Park left them in villages, but occasionally they were abandoned where they fell. On August 19, when the expedition arrived at Bamako on the Niger, only eleven British members had survived. Park and the remnants of his expedition hired canoes which took them downstream to Sansanding, a little eastward of Ségou, where they remained for two months in preparation for the passage downriver. Scott had died during the march, and Anderson died on

October 28. The expedition's survivors constructed a flat-bottomed boat from two native canoes which Park named HMS *Joliba*, the native name for the Niger. Only five of the British remained alive: Park, Lieutenant Martyn, and three soldiers. Isaaco was sent back with Park's final dispatches while the rest of the expedition sailed off down the Niger with many muskets and ample supplies.

In 1806, rumors about Park's death began to reach the coast. Isaaco was dispatched to the interior to find the truth, but all that he produced was Park's belt and a questionable account from Amadi Fatouma, who had guided Park downriver from Sansanding. Isaaco reported that Park had uncharacteristically shunned contact with the natives, offended the chiefs by refusing to pay their river customs, and fired upon anyone approaching the *Joliba*. Park sailed down the Niger past Timbukto to the village of Bussa (located in what would become Nigeria), where the natives attempted to stop his progress. During efforts to escape, the *Joliba* had capsized in the narrow Bussa Rapids, and Park and his companions had drowned. Although doubts about Park's death remained, later expeditions confirmed that he did die at the Bussa Rapids, but the manner of his death has always been subject to debate.

Impact

Mungo Park's second expedition was a tragic failure. Every European in his expedition perished, and despite the loss of life and the distance traversed, no new light had been cast on the termination of the Niger. Because of the uncertainty of distances, neither the coastward direction of the Niger nor the magnitude of Park's journey was immediately recognized. Park had commenced his last expedition erroneously believing that the Niger was the Congo River, and it is possible that he died holding that belief, despite having traveled more than three-fourths of the twenty-six-hundred-mile length of the river.

The supreme tragedy in the history of early African exploration was the loss of Park, one of the most respected explorers, in an expedition that added very little to geographical knowledge. His death was basically a result of two tragic errors in judgment: first, the decision to enter the bush country during the rainy season, and second, his avoidance of contact with the natives and his policy of firing on them. Park felt comfortable with the black Africans, but, by contrast, he feared the Moors. It must be remembered that Park left Sansanding a sick, desperate man who possibly lacked his normal clarity of judgment. Mungo Park created his own fame, and his achievements are remembered for the manner of his survival and for the death which made him and the Niger a single historical entity and inspired another generation of explorers.

Bibliography

Boahen, A. Adu. *Britain, the Sahara, and the Western Sudan, 1788-1861*. Oxford: Clarendon Press, 1964. General work on British exploration and trade in Africa. Boahen discusses Park's explorations in the context of British policy in Africa.

Brent, Peter. *Black Nile: Mungo Park and the Search for the Niger*. London: Gordon and Cremonesi, 1977. An excellent biography, well researched and handsomely illustrated.

Burns, Alan. *History of Nigeria*. London: Allen and Unwin, 1964. An excellent history of Nigeria with an emphasis on British influence. Includes a brief but valuable account of Park's explorations.

Gramont, Sandre de. *The Strong Brown God: The Story of the Niger River*. Boston: Houghton Mifflin Co., 1976. The best book on the European expeditions to the Niger River. Park's role and adventures are covered extensively and accurately.

Gwynn, Stephen L. *Mungo Park*. New York: G.P. Putnam's Sons, 1935. This is perhaps the best life of Park, but it is somewhat dated.

Langley, Michael. "The Last Journey of Mungo Park." *History Today* 21 (June, 1971): 426-432. A popular but well-written article on Park's fatal expedition of 1805-1806. Excellent illustrations and evaluation of Park's accomplishments.

Severin, Timothy. *The African Adventure*. New York: E.P. Dutton, 1973. A brilliant survey of precolonial expeditions in Africa. Contains new material and excellent illustrations. Good coverage of Park's life.

Phillip E. Koerper

ROBERT EDWIN PEARY

Born: May 6, 1856; Cresson, Pennsylvania
Died: February 20, 1920; Washington, D.C.

After several unsuccessful attempts, Peary became the first man to reach the geographic North Pole, on April 6, 1909.

Early Life

Robert Edwin Peary was born on May 6, 1856, in Cresson, Pennsylvania, a backwoods farm community. His New England forebears were Frenchmen (Peary is an American modification of the Gallic Pierre) who had made barrel staves for their livelihood. His father died when he was three and his mother, Mary Peary, was forced to rear her only child on meager resources.

His mother was extremely possessive and forced her son to dress in girlish clothes. Robert was nicknamed "Bertie" and he was regarded as a "sissy" by his peers. He would spend the remainder of his life attempting to compensate for his tortured early years.

Peary studied civil engineering at Bowdoin College in Brunswick, Maine, and resolved to outdo his rivals. He became active in sports, drama, and debate. Symbolically, he dressed up as Sir Lancelot at his college fraternity masquerade party. For graduation exercises, he composed an epic poem in which he imagined himself to be Sir Roland.

Life's Work

Peary received a degree in civil engineering in 1877 from Bowdoin College. After his graduation, he served as a draftsman for the United States Coast and Geodetic Survey. While in that position, he applied for and received a commission in the Civil Engineer Corps of the United States Navy in 1881.

In 1886, Peary borrowed five hundred dollars from his mother, took a summer leave of absence from the Navy, gathered a crew, and embarked on what would be the first of eight expeditions to the Arctic. Peary, along with a Danish skiing companion, made a one-

New York Times Co./Archive Photos

hundred-mile journey over the inland ice from the southwest coast of Greenland. The purpose of his first expedition was to acquire some fame by discovering what existed on north Greenland's ice cap: Was Greenland an island continent, or did it, as some geographers believed, thrust its ice cap right up to the North Pole? This expedition accomplished little. Yet Peary quickly learned what he needed to do in the future, and when his leave of absence expired, he returned to duty in Nicaragua with an obsession to return to the Arctic and to continue his quest for fame.

His second expedition was delayed until 1891. In 1888, he returned to his Navy job on the Nicaraguan canal route for what would be a two-year tour. That same year, Peary married Josephine Diebitsch, the daughter of a professor at the Smithsonian Institution. She was a tall, spirited woman whose appearance closely resembled his mother's. Peary's mother moved in with the newlyweds. This uncomfortable arrangement lasted a year. Josephine soon realized that her husband was really married to his Arctic adventures; to solve her dilemma, she accompanied him on his second expedition. By this point, Peary had become skillful in getting what he needed to continue his explorations. He pulled strings and used his gifted oratorical skills and enormous self-confidence to obtain ten thousand dollars from financial backers and an eighteen-month leave of absence from the Navy.

The stern, blue-eyed Peary sported a reddish-blond mustache; despite his serious nature, his overall appearance resembled that of the walruslike Ben Turpin, the silent-screen comedian. His face was already wrinkled from his time in Nicaragua and from exposure to Arctic blizzards and sun. His six-foot, sturdy physique, with broad shoulders and narrow hips, his finely tuned body which had already passed its thirty-fifth birthday, was ready for the mental and physical challenges ahead.

For his second trip to the Arctic, which began in 1891, his strategy was to take with him a party of six "campaigners," including Dr. Frederick Cook, and a seventh person, his wife, Josephine. Josephine attracted much attention from the newspapers: She would be the first white woman to winter at such a high latitude in Greenland. Once in position, Peary planned to conduct a "white march" over the great ice of northeast Greenland and to claim for the United States a highway to the North.

On June 6, 1891, the *Kite* sailed from Brooklyn, destined for the northwest coast of Greenland. Cook, nicknamed the Sigmund Freud of the Arctic, proved to be a helpful passenger; his obsession to reach the North Pole went back to his own deprived childhood, during which he won prizes in geography and worked in his free time to help support his poverty-stricken family. To pay for medical school, he had worked nights as a door-to-door milkman.

The *Kite* was in the process of ramming its way through the ice of Baffin Bay when Peary broke his lower right leg by striking it against the iron tiller. Cook quickly set the leg in splints, and Josephine relieved Peary's pain with morphine and whiskey. Peary would later praise Cook as a helpful and tireless worker who was patient and cool under pressure.

On July 30, 1891, the party landed on the foot of the cliffs in Inglefield Gulf, immediately north of Thule, the United States Greenland military base. With his right leg strapped to a plank, Peary continued to demonstrate leadership as he carried a tent ashore and supervised the construction of a prefabricated, two-room cabin named Red Cliff House. As the party settled in for the long polar night, the Etah Eskimos flocked from hundreds of miles away to see the first white woman to come to their country.

Peary soon began to recover from his broken leg. Josephine recorded in her journal that, within three months, he had discarded his crutches and had begun running foot races with Cook to build up his leg. The Eskimos watched as the white man took snow baths in subzero temperatures. To demonstrate his endurance to the Eskimos, he wore a hooded parka and caribou socks and slept in the open all night without a sleeping bag.

Peary realized that, to endure in the Arctic, he would have to adopt the survival techniques of the Eskimos. He learned from the Eskimos that expeditions required dog teams, sleighs, fatty meat for nourishment, and light fur garments. Yet he treated the Eskimos, who would continually come to his aid, as subhumans. He refused to learn their language, in contrast to Cook and Matthew Henson, and he rejected the hospitality of their igloos.

During May of 1892, Peary set out eastward, on his white march across the ice cap of north Greenland. Initially, the Eskimos and Cook accompanied him. Cook had gone ahead as a forward scout. When the two men rendezvoused, Peary ordered Cook to return to look after Josephine. The Eskimos feared that the evil spirit Tormarsuk presided in the interior, and they departed with Cook.

Peary and Eivind Astrup, a Norwegian ski champion, proceeded forward. In sixty-five days, the two men completed the unbelievable distance of six hundred miles over unknown terrain. On July 4, he named an easterly inlet Independence Bay, planted two American flags, and held a small celebration with the Norwegian skier. After they had rested, they turned around to retrace the six hundred miles back.

The trip would bring Peary fame; yet he had made costly cartographic errors that would eventually cause the death of a Danish scientist who attempted to confirm Peary's "discoveries." From Navy Cliff, Peary had believed that he had seen the Arctic Ocean, but he had actually been one hundred miles from the coast. Independence Bay had not been a bay but rather a deep fjord, and his conclusion that Peary Channel marked the northern boundary of the Greenland mainland was erroneous. In 1915, the United States government withdrew Peary's maps of Greenland, and Peary's reckless, unscientific behavior became legend.

When he returned, Peary raised twenty thousand dollars on the lecture circuit. Cook resigned from Peary's organization when Peary refused to allow him to publish ethnological findings on the Eskimos. No one in the group was allowed to publish anything, except in a book bearing Peary's name as author.

Peary's next expedition included Josephine, who was pregnant, a nurse, an artist, eight burros, and a flock of carrier pigeons. On September 12, 1893, Josephine gave birth to the first white child to be born at that latitude. The Peary's nine-pound daughter was named Marie Ahnighito Peary; her middle name came from the Eskimo woman who had chewed bird skins to make diapers for the blue-eyed child, nicknamed the Snow Baby.

The birth of the child was the only happy event of this expedition, as discontent broke out among the crew. Peary's drive and relentless nature began to cause problems. Astrup had a nervous breakdown and committed suicide on a glacier. Most of the remainder of the crew could no longer tolerate Peary, and they took the next supply ship back to the United States, as did Josephine, her child, and her nurse.

On future expeditions, Eskimos would lose their lives for Peary; Peary himself lost eight of his toes to frostbite. He stuffed his boots with tin-can lids to protect his stumps. Nothing short of death itself would stop Peary's single-minded quest.

Peary's first serious attempt to reach the North Pole began during the four-year expedition starting in 1898. Matthew Henson, an African American who had mastered the skills of Arctic exploration, was the only member from the original crew. The 1898-1902 mission failed to get to the North Pole, but Peary was able in 1902 to travel to eighty-four degrees, seventeen minutes north. On his seventh mission, in 1905-1906, Peary reached eighty-seven degrees, six minutes north, only 174 nautical miles from the North Pole, before having to retreat.

In 1908, Peary, though disabled, aging, and weatherbeaten, knew that he had the physical and mental resources for one final attempt to reach the North Pole. It would be his eighth and final trip. Several millionaires in the Peary Arctic Club pledged $350,000 for the final outing and *The New York Times* paid $4,000 in advance for the exclusive story. The National Geographic Society of Washington and the American Museum of Natural History in New York bestowed their prestige on him. The United States Navy once again released him with pay after President Theodore Roosevelt personally intervened.

In 1905, at a cost of $100,000, Peary had built, according to his own design, a schooner-rigged steamship named the *Roosevelt*. He took six men with him, the most loyal being Matthew Henson, who had nursed Peary and had saved his life on numerous occasions. A remarkable man, Henson had mastered everything for the mission, including the language of the Eskimos, who worshiped him as the Maktok Kabloonna (black white man).

The flag-decorated *Roosevelt* got under way from New York Harbor on the steamy afternoon of July 6, 1908. At Oyster Bay, Long

Island, President Roosevelt came aboard and shook hands with every member of the crew. At Sydney, Nova Scotia, Peary's wife Josephine, fourteen-year-old Marie, and five-year-old Robert, Jr., once again bade farewell to Peary. At Anoatok, on the northwest coast of Greenland, the Eskimos reported to Peary that Cook, who was involved in a rival expedition, had already passed westward on his march to the Big Nail.

Peary ordered Captain Bob Bartlett to begin ramming the *Roosevelt* through the ice packs and to head toward Cape Sheridan, the proposed wintering berth, which was 350 miles away. On September 5, the *Roosevelt* had reached her goal of eighty-two degrees, thirty minutes latitude—a record north for a ship under her own steam.

Cape Sheridan became home base. Ninety miles northwest lay Cape Columbia, which Peary decided would be the ideal jumping-off spot. Four hundred and thirteen miles of Arctic Ocean ice separated Cape Columbia from Peary's goal, the North Pole, the Big Nail, ninety degrees north latitude.

Peary was ready for the final chance to realize the greatest dream of his life. On the appointed Sunday morning, twenty-four men, 133 dogs, and nineteen sleighs departed. The expedition was broken up into five detachments: Each one would break trail, build igloos, and deposit supplies in rotation. Peary would follow the group from the rear as each exhausted team rotated back toward land.

On April 1, Bartlett took a navigational fix and determined a reading of eighty-seven degrees, forty-seven minutes north latitude. He took no longitudinal reading, which made his determination dubious, but Peary was convinced that he was 133 nautical miles on a direct beeline to the North Pole. Peary then surprised and disappointed Bartlett and ordered him home. The only qualified nautically trained witness who might verify the North Pole sighting finally departed.

Peary continued on with Henson, four Eskimos, five sleighs, and forty dogs. On April 6, 1909, at ten in the morning, after a labor of twenty years, Peary became the only white man to reach the North Pole. Once there, Peary draped himself in the American flag. Henson later recalled in his memoirs that his fifty-three-year-old commander was a dead-weight invalid, a mere shadow of the civil engineer in Nicaragua. One of the Eskimos remarked, "There is nothing here. Just ice!"

Impact

En route from his last mission to the Arctic, Peary learned that Cook had claimed to reach the Pole on April 21, 1908, nearly a year before Peary. Later, Cook was so hounded by the press and others that he took to wearing disguises and left the country for a year. When he returned, he spoke in his defense on the lecture circuit. Ultimately, his claims were disregarded, but the controversy was kept alive by the press because the dispute made a good story. Peary's claims were not scientifically documented. The National Geographic Society did a hasty, perfunctory examination of Peary's trunk of instruments in the middle of the night in a railway baggage station and agreed that Peary had discovered the North Pole.

In the 1930's, Dr. Gordon Hayes, an English geographer, scrupulously and fairly examined Cook's and Peary's claims. He concluded that neither one had gotten within one hundred miles of the North Pole. Nevertheless, Peary is credited with reaching the North Pole, attaining the fame he so desperately desired. After much lobbying and a congressional hearing, Peary was promoted to rear admiral. Peary served for a year as chairman of the National Committee on Coast Defense by Air during World War I. He retired and received a pension of sixty-five hundred dollars a year. He had achieved his goal, and the United States and the world recognized him for the twenty years of supreme sacrifices he had made.

Shortly after his return from the Arctic, Peary began suffering from anemia. On February 20, 1920, he died from that affliction at the age of sixty-four. He was buried with full honors at Arlington National Cemetery in Washington, D.C. His casket was draped with the remnants of the American flag with which he had covered himself as he stood atop the world on the North Pole. The National Geographic Society constructed a huge globe of white granite, representing the Earth and inscribed with Peary's motto, "I shall find a way or make one," and a legend proclaiming him Discoverer of the North Pole.

Bibliography

Bryce, Robert M. *Cook and Peary: The Polar Controversy, Resolved.* Mechanicsburg, Penn.: Stackpole Books, 1997. This lengthy text (1,133 pages) includes illustrations, maps, an extensive bibliography, and an index.

Cook, Frederick A. *My Attainment of the Pole.* New York: Polar, 1911. Cook's own descriptions of his expedition. Some claim that it was a hoax and others state that he only got to within a hundred miles of the North Pole.

Diebitsch-Peary, Josephine. *My Arctic Journal: A Year Among Ice-Fields and Eskimos.* New York: Contemporary Publishing Company, 1893. Peary's wife gives her account. Includes "The Great White Journey."

Henson, Matthew A. *A Black Explorer at the North Pole: An Autobiographical Report by the Negro Who Conquered the Top of the World with Admiral Robert E. Peary.* Foreword by Robert E. Peary. Introduction by Booker T. Washington. New York: Walker and Co., 1969. Henson's account. Henson began his life in poverty, attained fame, and ended his life as a parking attendant and a seventeen-dollar-a-week messenger in Brooklyn. He lived to be eighty-eight, and in his last years he suffered extreme poverty.

Herbert, Wally. *The Noose of Laurels: Robert E. Peary and the Race to the North Pole.* New York: Atheneum, 1989. Includes illustrations, a bibliography, and an index.

Hunt, William R. *To Stand at the Pole: The Dr. Cook-Admiral Peary North Pole Controversy.* New York: Stein and Day, 1982. Contains detailed account of the controversy over which man (Cook or Peary) got to the North Pole first. The mystery is not answered. Contains an excellent bibliography.

Peary, Robert E. *The North Pole.* New York: Stokes, 1910. Peary wrote three books—the others are *Nearest the Pole* (Doubleday, 1907) and *Northward over the "Great Ice"* (Stokes, 1898). *The North Pole* is Peary's own account of reaching the Pole. Exciting as an account but criticized by others.

Rasky, Frank. *Explorers of the North: The North Pole or Bust.* New York: McGraw-Hill Book Co., 1977. Chapters 10 and 11 are devoted to Peary. A human account of the explorer, warts and all.

Short, readable, extremely detailed report of the important events. Good starting point.

Rawlins, Dennis. *Peary at the North Pole: Fact or Fiction?* New York: Luce, 1973. Argues that Peary never made it to the North Pole.

Weems, John Edward. *Peary, the Explorer and the Man: Based on His Personal Papers*. Boston: Houghton Mifflin, 1967. New York: St. Martin's Press, 1988. Includes a foreword by Sir Edmund Hillary, illustrations, and a bibliography.

John Harty

WILLIAM PENN

Born: October 14, 1644; London, England
Died: July 30, 1718; Ruscombe, England

A leading Quaker, Penn contributed to the early development of the sect through his traveling ministry, his numerous religious tracts, intervention with English authorities for toleration, and establishment of Pennsylvania as a refuge for dissenters.

Early Life

William Penn was born October 14, 1644, on Tower Hill in London, England. His mother was Margaret Vanderschuren, the daughter of John Jasper, a Rotterdam merchant. His father, Sir William Penn, was an admiral in the British navy who first achieved prominence under Oliver Cromwell and, after the Restoration of the monarchy in 1660, went on to further success under the Stuarts. Despite some ups and downs in his career, the elder Penn accumulated estates in Ireland, rewards for his services, providing sufficient income so that the younger Penn was reared as a gentleman and exposed to the upper echelons of English society.

Penn received his early education at the Chigwell School, followed by a stint at home with a tutor who prepared him for entrance into college. In 1660, he was enrolled at Christ Church, Oxford University, where he remained until March, 1662, when he was expelled for infraction of the rules enforcing religious conformity. He then went on a grand tour of the Continent and spent a year or two at Saumur, France, studying languages and theology at a Huguenot school. After returning to England, he spent a year as a law student at Lincoln's Inn. His legal studies were somewhat sporadic and were never completed, a pattern that was typical for gentlemen of the day. They did, however, influence his subsequent writings and his ability to argue his own cause as well as those of others.

In Ireland, first as a child and later while acting as an agent for his father, Penn was exposed to Quakerism. It was in 1666, while managing the family estates, that he was converted by Thomas Loe, a Quaker preacher. Much to the horror of his father, young Penn took to preaching at Quaker meetings, quickly achieving prominence among the members of the still relatively new nonconforming sect. Parental disapproval continued until a reputedly dramatic reconciliation at the admiral's deathbed.

Life's Work
It was as a Quaker that Penn found his true calling. In 1668, he was in London preaching at meetings, and he produced his first religious tract, *Truth Exalted*. From that time onward, he spent a good portion of his life traveling, preaching, and writing religious tracts. He made several extended trips to the Continent, speaking at Quaker meetings as well as trying to convert others to the faith. By the end of his life, he had written some 150 works, most of them on religion. Some were descriptions of Quaker doctrines, such as *No Cross, No Crown* (1669); others were defenses of Quaker principles and actions—for example, *Quakerism: A New Nick-Name for Old Christianity* (1672).

There was no toleration in England in the 1660's for those who dissented from the Anglican Church. As a result, Penn, like numerous other Quakers, was arrested for attending Quaker meetings, for preaching, and for publishing a religious tract without a license. Indeed, several of his works were written while he was in prison. As a result of his experiences,

Archive Photos

as well as of observation of his coreligionists and friends, Penn became an advocate of religious toleration. He wrote several tracts, including *England's Present Interest Discovered* (1675), which pleaded with the government to recognize liberty of conscience for all, not only for Quakers. Penn used his position and friendship with the Stuart kings, Charles II and James II, to aid others.

Support for civil rights also came out of Penn's advocacy of religious toleration. His arrest for preaching at a Friends' meeting in 1670, a violation of England's stringent religious laws, led to two trials that ultimately contributed to the independence of juries. In the first case, Penn and his fellow Quaker William Mead were found not guilty of unlawful assembly by a jury which refused to alter its verdict after being ordered to do so or else "go without food or drink." Members of the jury were then fined; they appealed their case and ultimately were vindicated in their right to establish a verdict free from coercion.

Politically, Penn was caught, both in his beliefs and friendships, between the liberal dissenting Whig politicians and the conservative followers of the Stuart court. In 1678, he gave his support to Algernon Sidney's bid for a seat in Parliament, while maintaining his friendship with the Duke of York. Penn's attempt to keep his balance in the volatile English political scene of that period failed with the Glorious Revolution, when England exchanged the Catholic Duke of York, James II, for Mary, his Protestant daughter, and her husband, William of Orange. Accused of treason and at one point placed under arrest, Penn fled and for several years after 1691 went into hiding. It was not until after the turn of the century, under the rule of Queen Anne, that he again safely participated in the English political scene. Penn's position between the two major camps of the period is also evident in his writings and in the constitutional provisions he made for his colony, Pennsylvania, since they exhibit both liberal and conservative features.

For Americans, Penn is best known for his colonization efforts. His interest in the New World stemmed from his association with the great Quaker leader and preacher George Fox, who traveled through the Colonies. Penn's first involvement was in West Jersey, where he acted as an arbitrator in a complex dispute between two

Quakers with claims to that colony. He ultimately became one of the proprietors of West Jersey, as well as, after 1682, of East Jersey. Yet because Quaker claims to the government of the Jerseys were under question, he sought a colony of his own, and it was on Pennsylvania that he expended most of his efforts. In 1681, he obtained a charter from Charles II for extensive territories in America, ostensibly as payment owed to his father; the grant gave him rights to the government as well as the land of the colony.

In establishing Pennsylvania, Penn wanted both to create a refuge for Quakers where they would be free to worship without fear of imprisonment and a government with laws based upon their principles. At the same time, as proprietor of the colony, he hoped that the venture would be profitable. He started by preparing a constitution and laws for the colony, consulting numerous friends for their suggestions and comments. The resulting first Frame of Government proved too complex for the colony and was followed by other modified versions. Penn also worked to obtain both settlers and investors for his project and advertised it in pamphlets such as *A Brief Account of the Province of Pennsylvania* (1681) and *A Further Account of the Province of Pennsylvania and Its Improvements* (1685), which were published in several languages and distributed in both England and on the Continent. Expecting to be the resident proprietor and governor of the new colony, Penn made plans to move there. He journeyed to America twice, first in 1682 and again fifteen years later, in 1699, each time remaining for about two years. Both times he scurried back to England to protect his proprietorship, the first time from a controversy with Lord Baltimore, the proprietor of neighboring Maryland, over boundaries, and the second, to respond to a challenge from English authorities to all proprietary governments.

In the long run, Penn's colony was a success for everyone but him. His anticipated profits never materialized—a serious disappointment because, with advancing age, he was increasingly in financial difficulty. The Quaker settlers also proved to be a disappointment in their failure to get along with one another as well as with Penn. Indeed, they proved to be an exceedingly contentious lot, and the boundary controversy with Maryland was not solved in Penn's lifetime. After 1703, Penn negotiated with English authorities to sell his province back to the Crown, a deal which fell through because he suffered an incapacitating stroke in 1712. Pennsylvania, however, grew rapidly, and Philadelphia, the capital city that he had carefully planned, was an impressive success.

Penn is remembered for more than simply his religious writings and the establishment of Pennsylvania. In 1693, he wrote *An Essay Towards the Present and Future Peace of Europe*, which offered proposals for the establishment of peace between nations. In 1697, he proposed a plan of union for the Colonies, suggesting the creation of a congress of representatives from each colony which would meet once a year.

Penn was also a warm, affectionate, and concerned family man. In 1668, he married Gulielma Maria Springett; they had eight children, only three of whom survived childhood. In 1696, two years after his first wife's death, he married Hannah Callowhill, fathering another five children. Unfortunately, his children, like his colony, were a source of disappointment. His oldest son, and favorite, Springett, died at the age of twenty-one. His second son, William Penn, Jr., renounced Quakerism and was something of a rake. The surviving children of his second marriage were, at the time of his death, still young; it was to them that he left his colony of Pennsylvania.

Also contributing to Penn's woes in his later years was a festering problem with his financial agent, Philip Ford. Both were at fault, Ford for making inappropriate charges and Penn for a laxity in supervising his personal affairs. The result was that Ford's wife and children (after his death) pushed for payment—including Pennsylvania—for what they claimed were debts; they had Penn arrested and put in prison. When the dust settled, the Ford claims were taken care of and Pennsylvania had been mortgaged to a group of Penn's Quaker friends.

Summary

Although Penn was never more than a brief resident in the Colonies, his contributions to American history were substantial. He played a prominent role in the proprietorships of both East and West Jersey and was the founder of Pennsylvania. Penn was one of a handful of influential Quaker preachers and authors, and although his ideas were not original, he powerfully expressed and defended the sect's beliefs in numerous pamphlets, as well as in the laws and Frames of Government of Pennsylvania. As a colonizer, his efforts ensured a Quaker presence in America and the sect's role in the political and religious life of the middle colonies.

Penn's advocacy of religious toleration, of protection of the right to trial by jury, and of constitutional government carried across the Atlantic; as a result, provisions for all three were made in the colonies with which his name was associated. Penn thought that settlers would be attracted to America not only for its land but also for the freedoms it could offer, maintaining that Englishmen would only leave home if they could get more, rather than less, of both. He worked to make this happen. Thus, Penn used his connections among Whig and court groups on the English political scene to protect his fellow Quakers, his colony, and his proprietorship.

As the founder of Pennsylvania, he was the most successful of English proprietors and yet personally was a financial failure. He was a gentleman and a Quaker who could be contentious, particularly in religious debates, stubborn in maintaining his position against all opposition, and anything but humble in his life-style. In many ways, he was an uncommon and contradictory individual.

Penn's place in American history rests on his success in helping establish one colony and in founding another. The name Pennsylvania, standing for "Penn's woods," continues as a reminder of his significance. Sometimes overlooked, but also important, are his contributions to the fundamental political traditions which Americans have come to take for granted.

Bibliography

Beatty, Edward C. O. *William Penn as Social Philosopher*. New York: Columbia University Press, 1939. Reprint. New York: Octagon Books, 1975. Beatty examines Penn's philosophical and social ideas, viewing him as a political theorist, statesman, pacifist, humanitarian, and family man.

Bronner, Edwin B. *William Penn's Holy Experiment: The Founding of Pennsylvania, 1681-1701.* New York: Columbia University Press, 1962. Concerned with Penn's vision in establishing Pennsylvania and how it worked out. Contrasts plans and reality.

Dunn, Mary Maples. "The Personality of William Penn." *American Philosophical Society Proceedings* 127 (October, 1983): 316-321. Dunn portrays Penn as a restless rebel, a poor judge of people, and always the aristocrat.

_____. *William Penn: Politics and Conscience.* Princeton, N.J.: Princeton University Press, 1967. Argues that Penn was a creative thinker who, along with others of his age, wrestled with the question of what was constitutional government. The key to Penn's political ideas was liberty of conscience; a key to his behavior was the desire to protect his title to Pennsylvania.

Dunn, Richard S. "William Penn and the Selling of Pennsylvania, 1681-1685." *American Philosophical Society Proceedings* 127 (October, 1983): 322-329. Discusses Penn as a salesman and businessman.

Endy, Melvin B., Jr. *William Penn and Early Quakerism.* Princeton, N.J.: Princeton University Press, 1973. Concentrates on Penn's religious thought and its relationship to his political and social life. Also evaluates his significance for early Quakerism.

Illick, Joseph E. *William Penn the Politician: His Relations with the English Government.* Ithaca, N.Y.: Cornell University Press, 1965. Illick is primarily concerned with Penn as a practical politician who threaded his way through the perilous waters of English politics. Penn, as proprietor, is viewed as more successful, despite some ups and downs, in his dealings with the English government than with his own colonists.

Lurie, Maxine N. "William Penn: How Does He Rate as a Proprietor?" *Pennsylvania Magazine of History and Biography* (October, 1981): 393-417. Compares and contrasts Penn as proprietor with those who established other proprietary colonies.

Morgan, Edmund S. "The World and William Penn." *American Philosophical Society Proceedings* 127 (October, 1983): 291-315. A good, brief discussion of Penn's life which tries to put his experiences in the context of English society. Emphasis is on Penn as a Protestant, a gentleman, and an Englishman.

Morris, Kenneth R. "Theological Sources of William Penn's Concept of Religious Toleration." *Journal of Church and State* 35, no. 1 (Winter, 1993): 83-111.

Nash, Gary B. *Quakers and Politics: Pennsylvania, 1681-1726*. Princeton, N.J.: Princeton University Press, 1968. Concentrates on Penn's conflicts with the settlers in Pennsylvania, as well as their problems with one another. Nash emphasizes the religious and economic background of the disagreements. Good source for information on the dynamics of early Pennsylvania politics as well as on Penn.

Peare, Catherine Owens. *William Penn: A Biography*. Philadelphia: J. P. Lippincott Co., 1956. The standard modern work on Penn. A readable account with a sometimes excessively flowery style. There are numerous other, older, biographies, but this one is the best.

Penn, William. *The Papers of William Penn*. Edited by Mary Maples Dunn, Richard S. Dunn, et al. Philadelphia: University of Pennsylvania Press, 1981-   . This is the result of a project to collect and make available the widely scattered papers of Penn. Volumes are well annotated and a pleasure to use. In addition to the printed volumes, there is also a more complete microfilm series.

Maxine N. Lurie

ZEBULON MONTGOMERY PIKE

Born: January 5, 1779; Lamberton (Trenton), New Jersey
Died: April 27, 1813; York (Toronto), Ontario, Canada

Pike helped to open the American Southwest to U.S. interests, and he gave his name to the famous Pikes Peak in the state of Colorado.

Early Life
Zebulon Pike was born to a U.S. Army major and the former Isabella Brown. Pike was primarily raised on army posts in western Pennsylvania and Fort Washington, Ohio (now Cincinnati), while his father served under General James Wilkinson. Wilkinson, known to be wily, unscrupulous, and ambitious, would prove to be the predominant figure in the younger Pike's military and exploratory career.

Pike, a crack shot and expert outdoorsman, entered his father's company as a cadet when he turned twenty years of age but found himself relegated to the job of quartermastering supplies. Pike served on posts in Indiana and Illinois in essentially a peacetime army with no prospects for distinction until Wilkinson decided to use Pike in two expeditions exploring the opening American West. It is possible that Pike found favor with his father's former commander when he eloped with Clarissa Brown, a cousin on Pike's mother's side and the daughter of a Wilkinson family friend.

In 1805, after President Thomas Jefferson's Louisiana Purchase and the commission of the Meriwether LEWIS and William Clark expedition, the newly appointed Governor Wilkinson set up his headquarters in St. Louis, Missouri. There he prepared several expeditions to explore the upper Louisiana Territory, to establish the authority of the U.S. government over Indian tribes, and to discover the source of the Mississippi River. It is also possible that Wilkinson used Pike as leader of one of these expeditions to test

British reactions to American incursions into the fur-trapping country and further used the expedition to test Pike's capabilities for future commands.

On July 30, 1805, Lieutenant Pike received his orders; he left St. Louis on August 5 with a twenty-man party. Pike's team trekked two thousand miles by keelboat and on foot north to upper Minnesota. On September 23, Pike made a treaty with the Sioux Indians, purchasing one thousand acres of land that later became Minneapolis/St.Paul. On February 21 Pike reached Leech Lake, which he wrongly claimed was the source of the Mississippi. After wintering in the area, he returned on April 30, 1806. His tour was deemed successful, although Pike was disappointed that President Jefferson did not give him the accolades showered on Lewis and Clark's ongoing reports. Still, Wilkinson determined that Pike was dependable. Three days after Pike's return, Wilkinson told him he would command an especially important expedition into the Southwest.

Life's Work
While his motives and purposes remain uncertain, Wilkinson gave Pike command of a second expedition ostensibly to discover the headwaters of the Arkansas and Red Rivers and to explore them as far as the Spanish settlements of New Mexico in "New Spain." In addition, Pike was to accompany a party of Osage prisoners of war back to their country and effect a peace treaty between them and neighboring tribes, notably the strategically important Comanches. Wilkinson, a double agent for the Spanish court, cautioned Pike not to give the Spanish military cause for offense and not to reveal American interest in the economic, strategic, and settlement possibilities of the region. Oddly, Wilkinson did not provide Pike with enough provisions for the trip, did not assign enough men to repel a serious Indian attack, and did not obtain governmental authority for the expedition. Some historians believe Wilkinson planned to betray Pike to the Spanish in order to spark an international incident, although his own son was part of the company. It is possible Pike had an unspoken agreement with Wilkinson, went beyond his orders, or was an inept geographer; in any event, subsequent actions would cast this expedition under a cloud of mystery.

With a party of twenty-two men, including eighteen from his previous trip, Pike left on July 15, 1806, and marched west to the Arkansas River. Slowed by inclement weather, he moved through Pawnee villages on the Republic River in what is now Kansas. There he learned that six hundred Spanish troops were looking for him. Pike ignored Pawnee warnings to turn back and sent five men, including the younger Wilkinson, to follow the Arkansas River south and return to St. Louis to report. While not successful in his attempts to meet with Comanche chiefs, Pike was able to persuade the Pawnees that they should change their allegiance from Spanish to American interests.

After following Spanish trails en route to Santa Fe, on November 15, 1806, Pike saw Pikes Peak for the first time. Pike originally named the mountain "Grand Peak," but his cartographers labeled it "Pikes Peak" on their maps, despite the fact that the mountain was well known to the Spanish, Indians, and trappers who had previously crossed the region. On November 26, Pike made one attempt to climb the mountain but was unsuccessful given the cold winter storms; he likely climbed a nearby peak rather than the one that came to bear his name. Pike later predicted that no one would ever reach the summit of Pikes Peak. Pike then turned southwest, following Indian trails, and suffered from snow and ice in the Rocky mountains of Colorado and upper New Mexico. Pike then found himself traveling in circles, moving north instead of south and unable to determine which way to proceed. After leaving behind three men incapacitated by frostbite, Pike finally came to a halt when he believed he had again found the Red River.

After building a log fort to defend his party against Indians near what turned out to be the upper Rio Grande, on February 26 he was discovered by one hundred Spanish cavalry troops. They invited Pike to come to Santa Fe, where he and his men were held under arrest for illegal entry. Despite Spanish fears that Pike's and other excursions into their territory were evidence of future American invasions, Pike easily made friends with local authorities. Pike was treated well, but the Spanish confiscated most of his papers (some were hidden in his men's rifle barrels). Pike's party was then es-

PIKE EXPEDITIONS

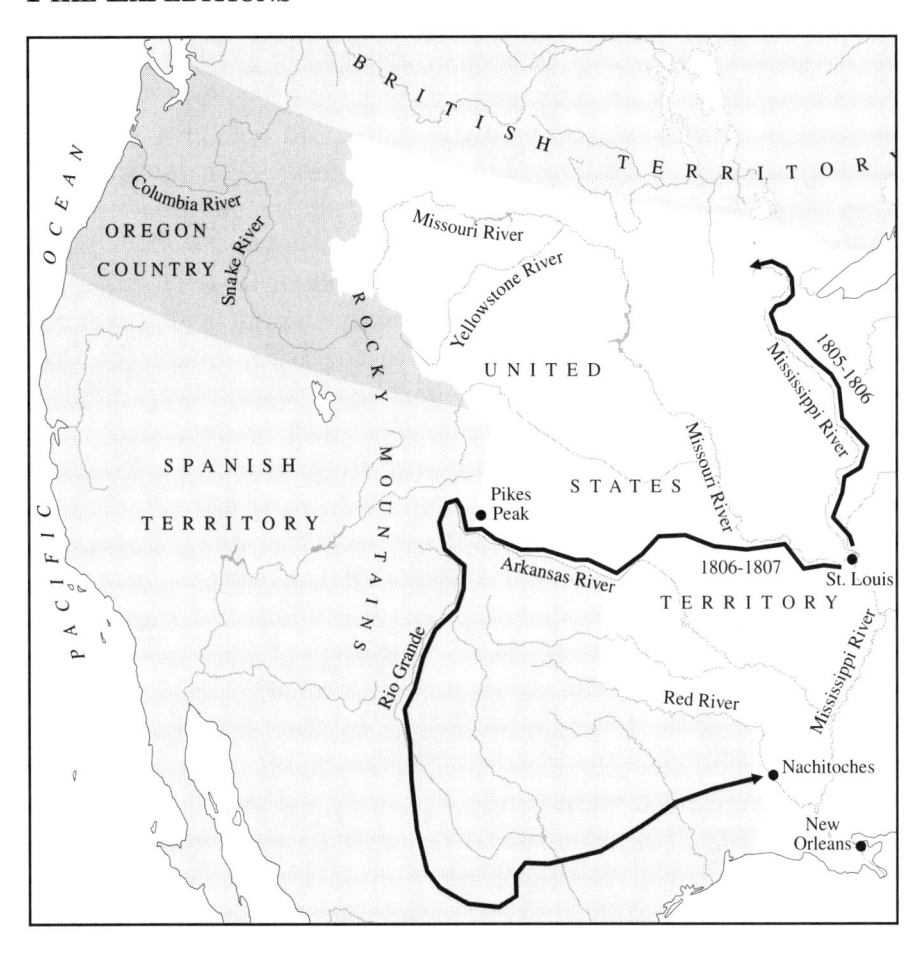

corted to Chihuahua, Mexico, and then Pike and his surviving men returned through Texas and were released in Natchitoches, Louisiana, on July 1, 1807.

Pike then went to Washington, D.C., but was busy with military business and did not present his report until January, 1808. By the time of his return, Wilkinson had been disgraced and Pike found that his own efforts were treated with indifference or contempt, particularly because he maintained Wilkinson's innocence in conspiring with former vice president Aaron Burr to create a new republic in the Southwest by seizing Spanish lands. Tainted by the

Wilkinson furor, Pike and his men were not given the usual land grants awarded by Congress to other explorers.

Pike's narrative of the trip was published in 1810. It presented the Great Plains to the American public for the first time as a Great American desert. Pike claimed the region did not have sufficient timber to support settlements, but he believed the desert would prove a useful buffer between the United States and Mexico. Furthermore, he claimed the plains would serve as an area to confine Indian tribes. Promoted to lieutenant colonel, in 1811 Pike was briefly under arrest for insubordination during the investigation of Wilkinson's possible involvement with the Burr conspiracy. Pike was completely exonerated on all charges. Wilkinson's role remains controversial and is still the subject of scholarly debate.

Becoming a general at the outset of the War of 1812, Pike was determined to find fame in combat or die. In April of 1813, Pike led sixteen hundred troops invading York (now Toronto), Canada, and was mortally wounded when the British deliberately set off a powder magazine. After a stone broke his spine, Pike lived long enough to hear the cries of his victorious men. As a gesture of respect, a captured British flag was placed under his head.

In 1908, Dr. H. E. Bolton discovered in Mexican archives some original papers confiscated during Pike's southwestern expedition of 1806. From these documents, Bolton and other historians concluded that Pike's second expedition was mounted primarily for the purpose of spying for Wilkinson and Aaron Burr, and only secondarily to gather topographical data. Later historians have refuted the charge, noting that extensive publicity preceded Pike's trip and the fact that Pike did not know Spanish. They also suggest that if Pike was part of the Burr conspiracy, he would logically have gone directly to the Southwest instead of following the Arkansas River.

Impact

Pike's western expedition pointed the way for new commercial interests, helped temporarily promote peace with the Indians, and helped establish the United States' domain over the territory. Pike's *An Account of Expeditions to the Sources of the Mississippi* (1810), although hastily compiled, inaccurate, and misleading, aroused

American curiosity about the West. His account of the weakness of Spanish authority in Santa Fe and the profitability of trading with Mexico stirred entrepreneurs and politicians to expand into Texas, setting the stage for the Mexican War. Some historians give Pike credit for helping to establish the myth of the Great American Desert, which retarded growth into the Great Plains. Pike's report also foreshadowed the later moves by the American government to push Indian tribes onto less desirable lands. Subsequent editions of Pike's journals were more dependable and more carefully organized, although not as honest or important as those of the LEWIS AND CLARK dispatches, from which Pike heavily cribbed his own data.

In 1820, Dr. Edwin James became the first man on record to climb Pikes Peak. Julia Archibald Holmes became the first woman to reach the summit in 1858. The same year, Pikes Peak or Bust became a familiar slogan painted on the wagons of gold miners attracted to the region. While few gold fortunes were made, the slogan made the mountain a national landmark. In 1893, Katherine Lee Bates, a Massachusetts author and teacher, composed the lyrics to "America the Beautiful," inspired by the view from the summit. The peak later became an important tourist attraction; visitors are able to reach the top by cog railway, car, or by foot. The Pikes Peak Cog Railroad was built in 1891 and is still in operation. Built in 1916, the Pikes Peak Toll Road is the second highest highway in the world. Pike County, Georgia, and its county seat, Zebulon, were named after Pike in 1856.

Bibliography

Carter, Carrol J. *Pike in Colorado*. Fort Collins, Colo.: Old Army Press, 1978. Carter's short, eighty-two-page volume focuses on Pike's discovery of Pikes Peak and his Colorado journey.

Hart, Stephen H., and Archer B. Hulbert, eds. *Zebulon Pike's Arkansas Journal*. Westport, Conn.: Greenwood Press, 1972. This edition focuses on Pike's lower Mississippi travels, augmenting Pike's accounts with interpretations based on recently discovered maps.

Hollon, W. Eugene. *The Lost Pathfinder: Zebulon Montgomery Pike*. Norman: University of Oklahoma Press, 1949. A longtime standard biography, Hollon's work is frequently cited as the most authoritative study of Pike's career.

Jackson, Donald, ed. *The Journals of Zebulon Pike, with Letters and Related Documents*. Norman: University of Oklahoma Press, 1966. Jackson's edition of Pike's journal is supplemented with both related documents and scholarly interpretations of Pike's accounts based on information discovered after Pike's death.

Pike, Zebulon Montgomery. *The Expeditions of Zebulon Montgomery Pike*. Edited by Elliott Coues. 2 vols. New York: Dover, 1987. Originally published in three volumes in 1895, these first-hand accounts of Pike's travels are accompanied by maps.

Sanford, William R., and Carl R. Green. *Zebulon Pike: Explorer of the Southwest*. Springfield, N.J.: Enslow, 1996. As part of the Legendary Heroes of the Wild West series, this biography is intended for the general reader.

Stallones, Jared. *Zebulon Pike and the Explorers of the American Southwest*. New York: Chelsea House, 1992. With an introductory essay by former Apollo astronaut Michael Collins, this short, critical overview of Pike emphasizes Pike's ineptitude and reliance on Wilkinson, portraying Pike as an unlucky man later mythologized as an American hero.

Terrell, John U. *Zebulon Pike: The Life and Times of an Adventurer*. New York: Weybright and Talley, 1968. Terrell is especially helpful discussing Pike's connections with his times, and color maps help illustrate Pike's journeys.

Wesley Britton

FRANCISCO PIZARRO

Born: c. 1495; Trujillo, Spain
Died: June 26, 1541; Lima, Peru

Pizarro was a sixteenth century Spanish conquistador who experienced many frustrating years in the New World in search of fame and fortune before discovering and conquering the Incan Empire of Peru.

Early Life

The details of Francisco Pizarro's early life are not clear. He was probably born around 1495 in Trujillo, a city in the province of Estremadura, Spain, from which came many of the famous conquistadores. Pizarro was one of several illegitimate sons of Gonzalo Pizarro, an infantry officer. His mother, Francisca Morales, was a woman of plebeian origin about whom little is known. He received little attention from his parents and was, apparently, abandoned in his early years. He could neither read nor write, so he became a swineherd and was so destitute that, like the prodigal son, he was reduced to eating the swill thrown out for the pigs. He probably needed little encouragement to abandon this ignoble profession to go to

Library of Congress

Seville, gateway to the New World and fame and fortune.

The circumstances under which Pizarro made his way across the Atlantic Ocean to the island of Hispaniola in the early years of the sixteenth century are not known. By then in his thirties, Pizarro was in his prime, yet his most productive years lay ahead. Contemporary portraits depict him as tall and well built with broad shoulders and the characteristic forked beard of the period. He possessed a noble countenance, was an expert swordsman, and had great physical strength. In 1510, he joined Alonso de Ojeda's expedition to Uraba in Terra Firma, where, at the new colony of San Sebastian, Pizarro gained knowledge of jungle warfare. When the colony foundered and Ojeda was forced to return for supplies to the islands, Pizarro was left in charge. He remained in the doomed colony for two months before death thinned the ranks sufficiently to allow the survivors to make their way back to civilization on the one remaining vessel. Shortly thereafter, Pizarro entered the service of Vasco Núñez de BALBOA and shared in the glory of founding a settlement at Darien and the subsequent discovery of the Pacific Ocean in 1513. Yet when Balboa fell from favor and was accused of treason by the governor of Panama, Pedrarias, Pizarro was the arresting officer. In the service of Pedrarias, there were new adventures, but at an age approaching fifty, old for that day, Pizarro had only a little land and a few Indians to show for his years of labor in the New World.

Life's Work

In 1515, Pizarro crossed the Isthmus and traded with the natives on the Pacific coast. There he probably heard tales of a mysterious land to the south rich beyond belief in gold and silver. The subsequent exploits of Hernán CORTÉS in 1519-1521 and an expedition by Pascual de Andagoya in 1522, which brought news of wealthy kingdoms, gave impetus to further exploration and greatly excited the cupidity of the Spaniards. To finance an expedition, Pizarro formed a business triumvirate with Diego de Almagro, a solider of fortune, and Hernando de Luque, a learned ecclesiastic.

Pizarro's first foray, launched in December, 1524, took him down the coast of modern Colombia, where he encountered every hardship imaginable and soon returned quietly to Panama. Under the guidance of Bartolomé Ruiz, a famous navigator and explorer, Pizarro's second expedition set sail in early 1526. The voyage took them beyond modern-day Ecuador into the waters south of the equator, where they found evidence of an advanced Indian civilization. An inadequate number of men, dwindling provisions, and hostile natives forced Pizarro and part of the company to take refuge first on the island of Gallo and later on Gorgona while Almagro returned to Panama to seek assistance. The governor, however, refused further help and sent a ship to collect the survivors. Audaciously, Pizarro and thirteen others refused to return. They endured seven months of starvation, foul weather, and ravenous insects until Almagro returned with provisions and the expedition was resumed. At length, they discovered the great and wealthy Incan city of Tumbes on the fringes of the Peruvian Empire. After a cordial stay with the natives, Pizarro returned to Panama with some gold, llamas, and Indians to gain support for an even greater expedition. The governor remained uninterested, so the business partners decided to send Pizarro to Spain to plead their case.

Charles V and his queen were sufficiently impressed with Pizarro's exploits and gifts to underwrite another expedition. In July, 1529, Pizarro was given extensive powers and privileges in the new lands, among them the titles of governor and captain-general with a generous salary. Almagro received substantially less, which caused a rift between the two friends. Before leaving Spain, Pizarro recruited his four brothers from Estremadura for the adventures ahead.

In January, 1531, Pizarro embarked on his third and last expedition to Peru. With no more than 180 men and three vessels, the expedition charted a course to Tumbes, which, because of a great civil war in the country, they found much less hospitable. Even so, the Spaniards' arrival was fortuitous in that the victor, Atahualpa, had not yet consolidated his conquests and was now recuperating at the ancient city of Cajamarca. In September, 1532, Pizarro began his march into the heart of the Incan Empire. After a difficult trek through the Andes, during which they encountered little resistance, they entered Cajamarca on November 15, 1532. Finding the Incan king at rest with only a portion of his army, Pizarro, pretending friendship, seized Atahualpa after a great slaughter of Indians. Atahualpa struck a bargain with his captors. In return for his release, he promised to fill a large room with gold. A second, smaller room was to be filled with silver. Fearing revolt, however, the captors carried out a summary trial, and the Inca was condemned to death.

Meanwhile, Almagro and his men had arrived in February, 1533, and loudly demanded a share of the wealth. The gold and silver vessels were melted down and distributed among the conquerors, while Almagro's men received a lesser amount and the promise of riches to come. Hernando Pizarro, Francisco's only legitimate brother, was sent to Spain with the royal one-fifth portion. From Cajamarca, Pizarro and his company pushed on to Cuzco. After encountering some resistance in the coutryside, the conquistadores entered the city on November 15, 1533, where the scenes of rapine were repeated again.

After the conquest of Cuzco, Pizarro settled down to consolidate and rule his new dominion, now given legitimacy and the name of New Castile in royal documents brought back from Spain by Hernando Pizarro. A new Inca, Manco Capac II, was placed on the throne, and a municipal government was organized after the fashion of those in Iberia. Most of Pizarro's time, however, was consumed with the founding of a new capital, Lima, which was closer to the coast and had greater economic potential. These were difficult years. In 1536, the Manco Capac grew tired of his ignominious status as a puppet emperor and led the Peruvians in a great revolt. For more than a year, the Incas besieged Cuzco. After great loss of life and much destruction throughout the country, the siege ended, although the Incas would remain restive for most of the sixteenth century.

In the meantime, a power struggle had developed between Almagro, who had returned from a fruitless expedition into New Toledo, the lands assigned him by the Crown, and Pizarro for control of Cuzco. On April 6, 1538, Almagro's forces were defeated in a great battle at Las Salinas. Almagro was condemned to death. In the three years that followed, Pizarro became something of a tyrant. On June 26, 1541, the Almagrists broke into Pizarro's palace in Lima and slew the venturesome conquistador.

Impact

There are, perhaps, two possible ways in which the career of Francisco Pizarro might be evaluated. On one hand, it is easy to regard him as one of many sixteenth century Spaniards, called conquistadores, whose cupidity sent them in search of fame and fortune, specifically gold and silver, in the New World. In a relatively short period of time, Incas everywhere were conquered, tortured, murdered, and systematically stripped of their lands, families, and provisions. Pizarro played a major role in the rapacious conduct of the Castilians. Although this view is not without some merit, it must be understood within the context of Pizarro's world. He was not unlike a medieval crusader who sallied forth against the enemy with the blessings of Crown and Church. The Crown was interested in precious metals and new territorial possessions, while the Church was concerned about lost souls. When his opportunity for fame and fortune finally presented itself, Pizarro had to overcome seemingly insurmountable odds—financial difficulties, hostile natives, harsh weather and terrain, and later the enmity of other conquistadores—to create a Spanish empire in South America. Although his methods cannot be condoned, the empires of Alexander the Great, Charlemagne, and other conquerors were fashioned in much the same way.

Bibliography

Birney, Hoffman. *Brothers of Doom: The Story of the Pizarros of Peru*. New York: G. P. Putnam's Sons, 1942. A well-written study of Pizarro and his brothers from the opening of the age of exploration to the death of Gonzalo Pizarro in 1548. The author purposely eschews footnotes and lengthy bibliographical references. A good introductory work.

Hemming, John. *The Conquest of the Incas*. New York: Harcourt Brace Jovanovich, 1970. A history of the conquest from BALBOA's discovery of the Pacific Ocean in 1513 through the disintegration of the Inca Empire, with reference to the life of Pizarro. Includes chronological and genealogical tables plus an excellent bibliography.

Howard, Cecil, and J. H. Perry. *Pizarro and the Conquest of Peru*.

New York: American Heritage, 1968. A well-illustrated history of the conquest and the civil wars which followed. Excellent for a younger reading audience.

Kirkpatrick, F. A. *The Spanish Conquistadores*. London: Adam & Charles Black, 1934, 2d ed. 1946. A survey of Spanish exploration, conquest, and settlement of the New World beginning with the voyages of Christopher Columbus. Provides a good overview of Pizarro's career.

Marrin, Albert. *Inca and Spaniard: Pizarro and the Conquest of Peru*. New York: Atheneum, 1989. Written for a juvenile audience, it describes how the Incan world was changed when Pizarro conquered Peru. Includes illustrations, maps, and a bibliography.

Means, Philip Ainsworth. *Fall of the Inca Empire and the Spanish Rule in Peru: 1530-1780*. New York: Charles Scribner's Sons, 1932. A history of the last years of the Inca Empire and Spanish dominion to 1780. Most of the important events of Pizarro's life are mentioned. Includes a scholarly bibliography plus a helpful index and glossary.

Prescott, William H. *The Conquest of Peru*. Revised with an introduction by Victor W. von Hagen. New York: New American Library, 1961. After more than a century and many editions, still one of the best works on the subject. Prescott's style will appeal to readers at all levels.

Varon Gabai, Rafael. *Francisco Pizarro and His Brothers: The Illusion of Power in Sixteenth-Century Peru*. Norman: University of Oklahoma Press, 1997. Translated from the Spanish by Javier Flores Espinoza. Includes illustrations, maps, and a bibliography.

Larry W. Usilton

MARCO POLO

Born: c. 1254; Venice?
Died: January 8, 1324; Venice

Through his Asian travels and his book recording them, Marco Polo encouraged a medieval period of intercultural communication, Western knowledge of other lands, and eventually the Western period of exploration and expansion.

Early Life

Despite his enduring fame, very little is known about the personal life of Marco Polo. It is known that he was born into a leading Venetian family of merchants. He also lived during a propitious time in world history, when the height of Venice's influence as a city-state coincided with the greatest extent of Mongol conquest of Asia. Ruled by Kublai Khan, the Mongol Empire stretched all the way from China to Russia and the Levant. The Mongol hordes also threatened other parts of Europe, particularly Poland and Hungary, inspiring fear everywhere by their bloodthirsty advances. Yet their ruthless methods brought a measure of stability to the lands they controlled, opening up trade routes such as the famous Silk Road. Eventually, the Mongols discovered that it was more profitable to collect tribute from people than to kill them outright, and this policy too stimulated trade.

Into this favorable atmosphere a number of European traders ventured, including the family of Marco Polo. The Polos had long-established ties in the Levant and around the Black Sea; for example, they owned property in Constantinople, and Marco's uncle, for whom he was named, had a home in Sudak in the Crimea. From Sudak, around 1260, another uncle, Maffeo, and Marco's father, Niccolò, made a trading visit into Mongol territory, the land of the Golden Horde (Russia), ruled by Berke Khan. While they were

Archive Photos

there, a war broke out between Berke and the khan of the Levant, blocking their return home. Thus Niccolò and Maffeo traveled deeper into Mongol territory, moving southeastward to Bukhara, which was ruled by a third khan. While waiting there, they met an emissary traveling farther eastward who invited them to accompany him to the court of the great khan, Kublai, in Cathay (modern

China). In Cathay, Kublai Khan gave the Polos a friendly reception, appointed them his emissaries to the pope, and ensured their safe travel back to Europe: They were to return to Cathay with one hundred learned men who could instruct the Mongols in the Christian religion and the liberal arts.

In 1269, Niccolò and Maffeo Polo finally arrived back in Venice, where Niccolò found that his wife had died during his absence. Their son, Marco, then about fifteen years old, had been only six or younger when his father left home; thus Marco was reared primarily by his mother and the extended Polo family—and the streets of Venice. After his mother's death, Marco had probably begun to think of himself as something of an orphan. Then his father and uncle suddenly reappeared, as if from the dead, after nine years of travel in far-off, romantic lands. These experiences were the formative influences on young Marco, and one can see their effects mirrored in his character: a combination of sensitivity and toughness, independence and loyalty, motivated by an eagerness for adventure, a love of stories, and a desire to please or impress.

Life's Work
In 1268, Pope Clement IV died, and a two- or three-year delay while another pope was being elected gave young Marco time to mature and to absorb the tales of his father and uncle. Marco was seventeen years old when he, his father, and his uncle finally set out for the court of Kublai Khan. They were accompanied not by one hundred wise men but by two Dominican friars, and the two good friars turned back at the first sign of adversity, another local war in the Levant. Aside from the pope's messages, the only spiritual gift Europe was able to furnish the great Kublai Khan was oil from the lamp burning at Jesus Christ's supposed tomb in Jerusalem. Yet, in a sense, young Marco, the only new person in the Polos' party, was himself a fitting representative of the spirit of European civilization on the eve of the Renaissance, and the lack of one hundred learned Europeans guaranteed that he would catch the eye of the khan, who was curious about "Latins."

On the way to the khan's court, Marco had the opportunity to complete his education. The journey took three and a half years by horseback through some of the world's most rugged terrain, including snowy mountain ranges, such as the Pamirs, and parching deserts, such as the Gobi. Marco and his party encountered such hazards as wild beasts and brigands; they also met with beautiful women, in whom young Marco took a special interest. The group traveled through numerous countries and cultures, noting the food, dress, and religions unique to each. In particular, under the khan's protection the Polos were able to observe a large portion of the Islamic world at close range, as few if any European Christians had. (Unfortunately, Marco's anti-Muslim prejudices, a European legacy of the Crusades, marred his observations.) By the time they reached the khan's court in Khanbalik (modern Peking), Marco had become a hardened traveler. He had also received a unique education and had been initiated into manhood.

Kublai Khan greeted the Polos warmly and invited them to stay on in his court. Here, if Marco's account is to be believed, the Polos became great favorites of the khan, and Kublai eventually made Marco one of his most trusted emissaries. On these points Marco has been accused of gross exaggeration, and the actual status of the Polos at the court of the khan is much disputed. If at first it appears unlikely that Kublai would make young Marco an emissary, upon examination this seems quite reasonable. For political reasons, the khan was in the habit of appointing foreigners to administer conquered lands, particularly China, where the tenacity of the Chinese bureaucracy was legendary (and eventually contributed to the breakup of the Mongol Empire). The khan could also observe for himself that young Marco was a good candidate: eager, sturdy, knowledgeable, well traveled, and apt (Marco quickly assimilated Mongol culture and became proficient in four languages, of which three were probably Mongol, Turkish, and Persian). Finally, Marco reported back so successfully from his first mission—informing the khan not only on business details but also on colorful customs and other interesting trivia—that his further appointment was confirmed. The journeys specifically mentioned in Marco's book, involving travel across China and a sea voyage to India, suggest that the khan did indeed trust him with some of the most difficult missions.

The Polos stayed on for seventeen years, another indication of how valued they were in the khan's court. Marco, his father, and his uncle not only survived—itself an achievement amid the political hazards of the time—but also prospered. Apparently, the elder Polos carried on their trading while Marco was performing his missions; yet seventeen years is a long time to trade without returning home to family and friends. According to Marco, because the khan held them in such high regard, he would not let them return home, but as the khan aged the Polos began to fear what would happen after his death. Finally an opportunity to leave presented itself when trusted emissaries were needed to accompany a Mongol princess on a wedding voyage by sea to Persia, where she was promised to the local khan. The Polos sailed from Cathay with a fleet of fourteen ships and a wedding party of six hundred people, not counting the sailors. Only a few members of the wedding entourage survived the journey of almost two years, but luckily the survivors included the Polos and the princess. Fortunately, too, the Polos duly delivered the princess not to the old khan of Persia, who had meanwhile died, but to his son.

From Persia, the Polos made their way back to Venice. They were robbed as soon as they got into Christian territory, but they still managed to reach home, in 1295, with plenty of rich goods. According to Giovanni Battista Ramusio, one of the early editors of Marco's book, the Polos strode into Venice looking like ragged Mongols. Having thought them dead, their relatives at first did not recognize them, then were astounded, and then were disgusted by their shabby appearance. Yet, according to Ramusio, the scorn changed to delight when the returned travelers invited everyone to a homecoming banquet, ripped apart their old clothes, and let all the hidden jewels clatter to the table.

The rest of the world might have learned little about the Polos' travels if fate had not intervened in Marco's life. In his early forties, Marco was not yet ready to settle down. Perhaps he was restless for further adventure, or perhaps he felt obliged to fulfill his civic duties to his native city-state. In any event, he became involved in naval warfare between the Venetians and their trading rivals, the Genoese, and was captured. In 1298, the great traveler across Asia and emissary of the khan found himself rotting in a prison in Genoa—an experience that could have ended tragically but instead took a lucky turn. In prison Marco met a man named Rustichello (or Rusticiano), from Pisa, who was a writer of romances. To pass the time, Marco dictated his observations about Asia to Rustichello, who, in writing them down, probably employed the Italianized Old French that was the language of his romances. (Old French had gained currency as the language of medieval romances during the Crusades.)

Their book was soon in circulation, since Marco remained in prison only a year or so, very likely gaining his freedom when the Venetians and Genoese made peace in 1299. After his prison experience, Marco was content to lead a quiet life in Venice with his family and bask in his almost instant literary fame. He married Donata Badoer, a member of the Venetian aristocracy, and they had three daughters—Fantina, Bellela, and Moreta—all of whom eventually grew up to marry nobles. Thus Marco seems to have spent the last part of his life moving in Venetian aristocratic circles. After living what was then a long life, Marco died in 1324, roughly seventy years of age. In his will he left most of his modest wealth to his three daughters, a legacy that included goods which he had brought back from Asia. His will also set free a Tartar slave, Peter, who had remained with him since his return from the court of the great khan.

Impact

The book that Marco Polo and Rustichello wrote in prison was titled *Divisament dou monde* (description of the world), although in Italian it is usually called *Il milione* (the million), and it is usually translated into English as *The Travels of Marco Polo*. The original title is more accurate than this English title, which is somewhat deceptive, since after its prologue the book is actually a cultural geography instead of a travelogue or an autobiography.

The book was immediately popular. Numerous copies were made and circulated (this was the age before printing), including translations into other dialects and languages. Some copyists were priests or monks who, threatened by descriptions of other religions and the great khan's notable religious tolerance, made discreet emendations. These changes may in part account for the emphasis on Christian miracles in the book's early sections and even for its anti-Muslim sentiments. The numerous manuscripts with their many variants have created a monumental textual problem for modern editors of the work, since Marco and Rustichello's original manuscript has disappeared.

Modern readers might be surprised by the book's impact in Marco's time and for centuries afterward, but to readers of the early fourteenth century, descriptions of Asia were as fantastic as descriptions of outer space are today. Unfortunately many people then tended to read it as though it were science fiction or fantasy, perhaps in part because of its romantic style (including Rustichello's embellishments). The title *Il milione*, whose origin is obscure, could refer to the number of lies the book supposedly contains. (Some readers considered Marco Polo merely a notorious liar.) Yet, allowing for textual uncertainties, modern commentators have judged the book to be remarkably accurate; thus, it was a valuable source for those readers who took it seriously. For centuries it was the main source of Western information about Asia, and it exercised a tremendous influence on the Western age of exploration (Christopher COLUMBUS carried a well-marked copy with him). It has also continued to influence the Western imagination—inspiring plays, novels, and films, as well as unrestrained scholarly speculation about Marco's life and travels. In short, Marco Polo has become a symbol of Western man venturing forth.

Yet in large part the meaning of Marco Polo's experience has been misinterpreted. His sojourn in the East has too often been seen as the first probe of Western man into unknown territory, with Marco as a kind of spy or intelligence gatherer identifying the locations of the richest spoils, the first example of Western man as conquerer (a viewpoint which is shamefully ethnocentric). While he did influence the Western age of exploration, conquest, and colonization, this was hardly his intent. Instead, Marco can best be seen as an exponent of intercultural communication who lived during a period when communication between East and West opened up for a brief time.

Bibliography

Calvino, Italo. *Invisible Cities*. Translated by William Weaver. New York: Harcourt Brace Jovanovich, 1974. Originally published in 1972 as *Le città invisibili*, this postmodernist novel by one of Italy's leading writers is a fascinating example of an imaginative work inspired by Marco Polo. Consists of conversations between Marco Polo and Kublai Khan and Marco's descriptions of imaginary cities.

Li Man Kin. *Marco Polo in China*. Hong Kong: Kingsway International Publications, 1981. Although poorly written and edited, this volume is a good example of a freely speculative work about Marco from a non-Western point of view. Includes good illustrations, although they are not closely related to text.

Olschki, Leonardo. *Marco Polo's Asia: An Introduction to His "Description of the World" Called "Il milione."* Translated by John A. Scott. Berkeley: University of California Press, 1960. The best scholarly introduction to Marco Polo and his book. Discusses in detail the book's treatment of such topics as nature, politics, religion, Asian history, historical and legendary figures, and medicine.

Polo, Marco. *The Travels of Marco Polo*. Translated by Ronald Latham. Harmondsworth, England: Penguin Books, 1958. The best translation into English of Marco's book. Based on modern textual scholarship. Contains a brief but good introduction by the translator.

Power, Eileen. "Marco Polo: A Venetian Traveler of the Thirteenth Century." In *Medieval People*. 10th ed. New York: Barnes and Noble Books, 1963. A colorfully written account of Marco's travels, with descriptions of cities and rulers and quotations from other European travelers of the time who visited Asia.

Stefoff, Rebecca. *Marco Polo and the Medieval Explorers*. New York: Chelsea House Publishers, 1992. Part of Chelsea's juvenile World Explorers series. Includes illustrations and a bibliography.

Wood, Frances. *Did Marco Polo Go to China?* London: Secker & Warburg, 1995. Boulder, Colo.: Westview Press, 1996.

Harold Branam

JUAN PONCE DE LEÓN

Born: 1460; Tierra de Campos, Valladolid Province, Spain
Died: 1521; Havana, Cuba

Explorer of the "New World," Ponce de León founded the oldest settlement in Puerto Rico and discovered Florida while searching for the mythical Fountain of Youth.

Early Life

Juan Ponce de León was born in the province of Valladolid in Spain in 1460 and was a member of a noble family whose history is connected with the old kingdom of León. As a young boy, he served as a page to Pedro Núñez de Guzmán, Lord of Taral, and received military instruction from some of the most brilliant officers of that period. He learned tactics for outmaneuvering the opposition even when he was operating against forces with much larger numbers than his own. As a teenager, he fought in many campaigns against the Moors in Granada, learning the type of warfare that benefited him later in battles with the Indians in the New World. Ponce de León was an excellent judge of character and selected high-quality men for the important adventures and duties he was called on to perform. In the field, he was a superior leader, possessed of a vigorous, aggressive constitution, and was known as one of the bravest Spanish soldiers, regarding the safety and welfare of his men as more important than his own.

When Christopher COLUMBUS outfitted his second expedition to the New World in 1493, Ponce de León was one of the first to volunteer his services. After arriving in Cuba, Ponce de León was assigned next in command to the Spanish military leader Juan de Esquival. Between 1502 and 1504, Ponce de León led several successful military campaigns against the Indians in Higüey, the eastern province of Hispaniola (now Haiti and the Dominican Republic).

As a reward, he was appointed the governor of Higüey under the tutelage of Nicolás de Ovando, governor of Hispaniola.

From the Indians who visited Higüey frequently, Ponce de León learned that the rivers and mountains of the nearby island of Borinquén (now Puerto Rico) contained large quantities of gold. Receiving permission from Governor Ovando, Ponce de León set out to explore Borinquén in 1508. With twenty men, Ponce de León landed once more on the same beach he had explored with Columbus fifteen years earlier. He led his company to the main Indian settlement of the island and befriended the head chief of the Arawak tribe, Agüeybana, who soon led Ponce de León to the Manataubon and Zebuco Rivers, where rich deposits of gold nuggets were found by Ponce de León and his men. With the help of Agüeybana, Ponce de León built a small stone fortress, two farmhouses, and a short but adequate dock. He then returned to Hispaniola and recruited more Spanish settlers to join him in Puerto Rico, where he established a new Spanish colony known as Caparra.

Life's Work

In 1509, when Governor Ovando was called back to Spain, he gave a very favorable report of the conduct and leadership of Ponce de León and the merits of his services to the Crown. Consequently, King Ferdinand of Spain appointed Ponce de León to serve as governor of Puerto Rico. After firmly establishing his government, Ponce de León divided Puerto Rico into districts and towns and distributed the Indians into various groups. Ponce de León hoped to treat the Indians fairly, but the pressure for laborers was too strong. To the Spanish settlers, it seemed only natural to require the Indians to work for them, and so the Indians were treated as slaves.

The Indians soon rebelled against the heavy tasks imposed upon them and staged an uprising. However, even though he was greatly outnumbered, Ponce de León used his military expertise to defeat the Indians. Afterward, the Indians of Puerto Rico never combined in war against Ponce de León again, but they dispersed into the forests and mountains, where they were employed laboring in the mines, growing grain, cotton, and sugarcane, and performing other arduous tasks. In time, the Indians disappeared as a separate group of people in Puerto Rico. Some of them escaped from the island to nearby islands, while others intermarried with the Spaniards. In 1511, Ponce de León imported the first native Africans into Puerto Rico to perform slave labor. Ponce de León became one of the wealthiest and most powerful Spaniards in the New World.

While Ponce de León had been engaged fighting the Indians to maintain his control in Puerto Rico, King Ferdinand replaced him as governor in 1512 with political rival Juan Cerón. King Ferdinand ordered Cerón to treat Ponce de León with respect and honor, cultivate friendship with him, and not confiscate any of his land, houses, or riches. Intrigued by Cerón's wealth, Ponce de León sought adventures that would yield such riches, becoming enthralled with the idea of finding the mythical Fountain of Youth as well as more gold. King Ferdinand granted him permission to find and colonize the island of Bimini, the supposed site of the Fountain of Youth. According to medieval folklore, the fountain's water was the Water of Life in the Garden of Eden. The Caribbean Indians described it as a rejuvenating, tonic spring that existed somewhere north of Puerto Rico; it supposedly restored youth to old people who bathed in it or drank its waters.

In 1513, Ponce de León prepared three ships at his own expense and led an expedition in search of Bimini. After twenty-five days of sailing among the Bahamas, visiting several islands that were yet unknown to Europeans, Ponce de León reached the coast of present-day Florida on April 3, 1513. During the next few days, he led his expedition ashore, somewhere between St. Augustine and the St. Johns River. Thinking that this was another Caribbean island, he

Library of Congress

claimed the land for Spain in the name of King Ferdinand. According to legend, since Ponce de León arrived on the new land during the Easter season, when many flowers were profusely blooming, he named the land Pascua Florida, meaning the Feast of Flowers at Easter time. Although there exists some evidence for the prior discovery of Florida by John and Sebastian Cabot, the official credit for finding Florida goes to Ponce de León.

Later in April, 1513, Ponce de León sailed northward as far as the mouth of the St. Johns River and then turned southward, stopping at Cape Canaveral and at Biscayne Bay. Passing on down the line of the Florida Keys and the Tortugas, he turned northward and proceeded up Florida's west coast as far as Charlotte Harbor and possibly to Pensacola Bay, after which he returned to Puerto Rico in September. On his journey back to Puerto Rico, he landed on what is now Mexico's Yucatán Peninsula and also discovered the Bahama Channel, the seaway for the Spanish fleet to move from the Caribbean into the Atlantic. Although Ponce de León was disappointed that he had not found the imaginary island of Bimini, he had discovered Florida and explored a part of the North American mainland.

In 1514, Ponce de León sailed to Spain to report his recent discoveries, and King Ferdinand directed him to return to the New World and colonize both Florida and Yucatán, as well as to rid the West Indies of the Carib Indians, who were fierce cannibals. From 1515 to 1521, he was engaged in subduing the Caribs and other rebellious tribes on Puerto Rico and nearby islands, as well as devoting himself to personal and family matters. During this period, he was also influential in building up the city of San Juan, the current capital of Puerto Rico, and spent some time occupying Trinidad.

Finally, in February of 1521 Ponce de León was able to make his long-awaited second voyage to Florida in order to establish a Spanish colony. He took two ships, two hundred colonists, fifty horses, livestock, cats to catch mice in granaries, crossbows, spears, guns, farm implements, and both secular and monastic clergy. The expedition reached the west coast of Florida at Charlotte Harbor, went ashore, and began the construction of shelters and houses. Fearing the invasion and possible subjugation to the Spaniards, a large party of Apalachee Indians attacked Ponce de León's colony and drove them back to their landing site. Ponce de León organized his men into military groups and made a determined charge on the Indians, completely breaking up the Indian attack. However, while reforming his troops for another attack, a concealed Indian fired an arrow into Ponce de León's thigh close to the femoral artery. He was carried on board his ship, where he ordered his men to sail to the nearest island, Cuba.

The Indian's arrow had broken off in Ponce de León's bone, and the surgeons were unable to remove it. Soon after Ponce de León arrived in Havana, gangrene developed, and he died. He was buried in Havana with full military honors. Later in the century, his remains were removed to San Juan, Puerto Rico, to rest in the country to which he had devoted much of his life and energy. A marble tomb in the San Juan Cathedral holds the remains of Ponce de León, and the epitaph inscribed on his tomb reads "In this sepulcher rest the bones of a man who was a lion by name and still more by nature."

Impact

Juan Ponce de León was one of Spain's most gallant and faithful explorers and colonizers. He was an energetic, influential leader who inspired those around him to excel. Between 1508 and 1509, he established the first settlement in Puerto Rico, serving as the first governor of the island from 1509 to 1512. This small green island found by Christopher COLUMBUS and settled by Ponce de León grew to be an important Spanish colony and the key to Spanish defenses in the Caribbean.

Ponce de León discovered many other Caribbean islands in his search for the island of Bimini. While exploring for the mythical Fountain of Youth, he became one of the first Europeans to explore what is now the United States, and although he never discovered the Fountain of Youth, he was the first explorer to claim part of the North American mainland for Spain. He also discovered the Bahama Channel and was the first to report the strength of the Gulf Stream in propelling ships from the Caribbean into the Atlantic. His discovery of Florida in April, 1513, was one of the primary events in the colonial expansion of Europe that led other explorers to search for new trade routes and to explore the Western Hemisphere at the end of the fifteenth century.

Bibliography

Baker, Nina Brown. *Juan Ponce de León*. New York: Alfred A. Knopf, 1957. A bibliographical account of the explorations of Ponce de León from his days with Columbus to his death in Cuba.

Blassingame, Wyatt. *Ponce de León: A World Explorer*. Champaign, Ill.: Garrard, 1965. A biography of the Spanish explorer that concentrates on his travels in the Caribbean, his discovery of Florida, and his lifelong quest for the Fountain of Youth.

Brau, Maria M. *Island in the Crossroads: The History of Puerto Rico*. Garden City, N.Y.: Doubleday, 1968. This work traces the history of Puerto Rico from the first Spanish settlement established by Ponce de León to the 1960's.

Brown, George M. *Ponce de León Land and Florida*. St. Augustine, Fla.: Record Printing Company, 1952. A detailed description of the adventurous activities of Ponce de León in Puerto Rico and Florida, including some insights into his love life.

Jahoda, Gloria. *Florida*. New York: W. W. Norton, 1976. Documents Ponce de León's discovery and exploration of Florida, as well as his death.

McGuire, Edna. *Puerto Rico*. New York: Macmillan, 1963. This account of the history of Puerto Rico gives an excellent description of Ponce de León's life in Puerto Rico, his governorship, and his interaction with the Indians.

McMurry, Charles A. *America Discovery and Exploration*. New York: Macmillan, 1904. An account of six famous explorers of the New World is presented, one being Ponce de León. Highlights his explorations of and contributions to the New World.

Tebeau, Charlton W. *A History of Florida*. Coral Gables, Fla.: University of Miami Press, 1971. Chapter 2 of this book deals with the exploration of Florida and presents a good synopsis of Ponce de León's life, particularly his discovery and exploration of Florida.

Alvin K. Benson

JOHN WESLEY POWELL

Born: March 24, 1834; Mount Morris, New York
Died: September 23, 1902; Haven, Maine

In 1869, Powell led the first party of exploration to descend the gorges of the Green and Colorado Rivers by boat, stimulating interest in the geology and scenic wonders of the Grand Canyon. He also helped to establish the concepts of large-scale damming and irrigation projects as the keys to settlement and agricultural survival in the arid lands of the American West beyond the one hundredth meridian.

Early Life

John Wesley Powell was born on March 24, 1834, at Mount Morris, New York, the son of a circuit-riding Methodist minister who supplemented his income by farming and tailoring. The family moved to Ohio in 1841. The abolitionist views of the Powell family were not well received in Ohio, and John Wesley had such a difficult time at school that he was eventually placed under the direction of a private schoolmaster. This proved a significant experience, for the young Powell accompanied his tutor on biological field trips and developed a strong interest in both biological and physical science. The family eventually moved on to Illinois, where John Wesley grew to maturity. He spent several years combining a career as a teacher in Wisconsin and Illinois with sporadic attendance at several colleges, including Wheaton, Oberlin, and Illinois College. During this period he undertook extensive natural history excursions and ambitious journeys by boat down the Illinois, Des Moines, Ohio, and Mississippi Rivers from St. Paul all the way to New Orleans.

When the Civil War came, Powell immediately enlisted as a private in an Illinois volunteer infantry company. He rose quickly through the ranks and became a student of military engineering and fortifications. He met and became a friend of General Ulysses S.

Archive Photos

Grant and eventually commander of his own battery in an Illinois artillery unit. He led his battery into the fierce struggle at the Hornet's Nest in the Battle of Shiloh, where he was hit by a Minié ball, requiring the amputation of his right arm. Despite his injury he continued in service, seeing action and carrying out important duties in a number of major campaigns and rising to the rank of brevet lieutenant colonel.

After the war, Powell returned to Illinois and became professor of natural history at Illinois Wesleyan College, later moving to Illinois Normal University at Bloomington. By this time he had become accustomed to taking his students into the field as part of their training, but he was increasingly obsessed with the desire to reach farther afield. He was particularly drawn by the glamour and mystery of the trans-Mississippi West and began to assemble the ingredients that would allow him to make his first major expedition into that area.

Powell was instrumental in the establishment of a state natural history museum in Bloomington, and as its first curator he secured funding from several governmental and private sources to undertake a collecting expedition into the West. His friendship with General Grant enabled him to arrange for low-cost rations from army posts and for military protection for part of his trip. In 1867, the expedition, including students, amateur naturalists, teachers, and family members, set out from Council Bluffs on the first of Powell's major expeditions. The summer was spent examining the country and collecting specimens in the Colorado Rockies, and Powell remained after most of his party returned east and journeyed along the Grand River in Colorado.

The following summer, Powell returned to the Rockies with an expedition of twenty-five people, sponsorship from various Illinois State institutions, and encouragement from officials of the Smithsonian Institution, who were intrigued by his plans to explore among the rivers and high peaks of Colorado. After time collecting specimens in the Middle Park region, in late August Powell and six of his party made the first ascent of Long's Peak. They then moved into the White River basin, intending to follow it down to the Green River and on to a winter reconnaissance of the Colorado River. By now Powell had become thoroughly captivated by Western adventuring and scientific exploration and was obsessed by the unknown mysteries and legends of the Colorado. He had actively promoted his ideas and successfully publicized his activities and plans and had something of a reputation as an explorer and scientist, as well as good connections in the political and scientific communities. Although this five-foot-six, bearded veteran with only one arm hardly looked the part of the great explorer, Major John Wesley Powell was on the threshold of one of the great Western adventures.

Life's Work
The gorges of the Green and Colorado Rivers were among the few remaining unexplored areas on the North American continent. The legends which had been constructed out of the tales of Indians, mountain men, and other sources told of a region of enormous waterfalls, vicious whirlpools and rapids, and enormous rock cliffs which offered no escape or refuge from the punishment of the river. Essentially Powell and his men would plunge into a river descent of nearly nine hundred miles with no real idea of what terrors and adventures lay before them. Back east, Powell made the best preparations he could. A Chicagoan built four small wooden boats, one sixteen feet long of pine, the other three twenty-one feet, of oak, with water-tight compartments. Powell secured some financial support from a variety of public and private sources, although most of the meager financing came out of his own pocket. He assembled a varied group of nine companions, and on May 24, 1869, after several weeks of training, they set off down the Green River toward the Colorado.

They were on the river for ninety-two days. Their small vessels plunged through turbulent rapids, foaming cataracts, and towering canyon walls that at least matched most of the myths and legends. Two boats were lost, one expeditioner deserted early, three others were killed by Indians as they gave up on the river journey and attempted to climb out of the Grand Canyon. A confidence man surfaced who claimed that he was the only survivor of a wreck beneath a falls that had claimed the lives of the other members of the expedition, and newspapers across the country reported that Powell's party had been defeated by the river. By the time they in fact surfaced at a Mormon settlement below the canyon, Powell and his men had explored the Colorado River and the Grand Canyon and had discovered the last unknown river and mountain range in the American West. Powell's prodigious expedition marked him immediately as an American hero and one of the great explorers in the nation's history. It also meant that he could attract support and financing for further activities.

Powell returned to the Colorado two years later and retraced his original steps, now with the sponsorship of the Smithsonian Institution and the Department of the Interior. This expedition was a more determinedly scientific endeavor, operating as a survey group, the United States Geological and Geographical Survey of the Rocky Mountains, and they undertook a careful study, survey, and mapping of the canyon country. Powell became fascinated with the question not only of how the region—its canyons, plateaus, and mountains—looked but also of how they had been formed. He undertook additional Western expeditions and employed men who explored the high plateaus of Utah, the Colorado Plateau, Zion and Bryce Canyons, and the Henry and Uinta Mountains. The work of Powell and his associates introduced the idea of vast processes of uplift and erosion as responsible for the topography of the canyon and plateau country. They helped to popularize the geological concept of "base level of erosion." Powell's findings and ideas were published as *Explorations of the Colorado River of the West and Its Tributaries* (1875; revised and enlarged in 1895 under the title *Canyons of the Colorado*).

Powell's interest in the topography and geology of the Western regions led him naturally to a concern about the management of its lands. In 1878 he published *A Report on the Lands of the Arid Region of the United States*, which has been described as among the most important works ever produced by an American. Powell rejected both the concept of the inexhaustibility of natural resources and the idea that the West was the "Great American Desert" and not capable of supporting substantial settlement. Powell's familiarity with the West had convinced him that its lands and climate west of the one hundredth meridian were simply not suitable for development under policies that had been shaped by the conditions in the eastern regions. The arid lands of the West required a different strategy, and the key was water management.

Powell argued that the arid regions would not support the traditional family farm on the eastern model and that the lands of the

West should be categorized and utilized according to their most efficient uses for grazing, lumbering, mining, farming, and other purposes. Water should be considered a precious resource to be allocated by the community for the benefit of society in general rather than a privileged few. Government should undertake large-scale damming and irrigation projects so that the arid regions could be "reclaimed" and become productive. Powell's ideas represented a significant departure from the conventional wisdom regarding land use and the West, and his prestige as an explorer and scientist, coupled with his office as director of the United States Geological Survey from 1881 to 1894, put him in a position to be enormously influential in shaping the establishment in 1902 of the United States Bureau of Reclamation, which helped to make water management one of the major components of the early conservation movement.

During his Western expeditions, Powell had become fascinated by the cultures of the Indian tribes of the region, and it is characteristic of the man that he became a student of anthropology and headed the Bureau of Ethnology of the Smithsonian Institution during the same period that he led the Geological Survey. In 1880 he published his *Introduction to the Study of Indian Languages*.

Major Powell's retirement in 1894 was brought about partially because of physical ailments and partially because of his frustration in trying to get his ideas implemented. Ironically, his death in 1902 coincided with the passage of the Reclamation Act, which institutionalized many of his theories concerning land and water management.

Impact

Powell's career was significant on several fronts. As an explorer, his journey down the Colorado River through the Grand Canyon in 1869 ranks as one of the epic American adventures. His scientific background and interests prepared him for important accomplishments in mapping, surveying, and studying the geology of the plateau and canyon country, and for long service as director of the United States Geological Survey. During the same period, he headed the Bureau of Ethnology of the Smithsonian Institution. Powell became most interested in the problems of proper management and utilization of the lands in the arid West and was convinced that intelligent water management was the key to its development. He is one of the fathers of the concept of "reclamation" of arid lands through the construction of dams and irrigation projects.

Bibliography

Aton, James M. *John Wesley Powell*. Boise, Idaho: Boise State University, 1994. This brief biography (55 pages) is no. 114 in the Boise State University Western Writers Series.

Bartlett, Richard A. *Great Surveys of the American West*. Norman: University of Oklahoma Press, 1962. A comprehensive treatment that includes the work of Powell.

Cooley, John, ed. and comp. *The Great Unknown: The Journals of the Historic First Expedition Down the Colorado River*. Flagstaff, Ariz.: Northland Publishers, 1988. Includes a bibliography and an index.

Darrah, William C. *Powell of the Colorado*. Princeton, N.J.: Princeton University Press, 1951. A useful scholarly biography. Well researched, drawing on some unpublished sources, but rather colorless. Includes illustrations.

Exploring the American West, 1803-1879 (National Park Handbook no. 116). Washington, D.C.: Government Printing Office, 1982. This 128-page booklet is profusely illustrated and contains several photographs of Powell and his survey. The text is by William H. Goetzmann.

Fradkin, Philip L. *A River, No More: The Colorado River and the West*. New York: Alfred A. Knopf, 1981. Focusing upon the Colorado River and its tributaries, Fradkin discusses the federal land and water policies that shaped much of the West. Powell's role in the evolution of these developments is considered.

Gaines, Ann. *John Wesley Powell and the Great Surveys of the American West*. New York: Chelsea House Publishers, 1991. This juvenile biography is part of Chelsea's World Explorers series and includes illustrations and a bibliography.

Goetzmann, William H. *Exploration and Empire: The Explorer and the Scientist in the Winning of the American West*. New York:

Alfred A. Knopf, 1966. This Pulitzer Prize-winning book is the standard general treatment of the role of exploration in the American West. Contains a chapter dealing with Powell's life and career.

Savage, Henry, Jr. *Discovering America, 1700-1875*. New York: Harper and Row, Publishers, 1979. A very readable survey which is particularly good on the nineteenth century explorations.

Schwartz, Seymour I., and Ralph E. Ehrenberg. *The Mapping of America*. New York: Harry N. Abrams, 1980. An enormously detailed and lavishly illustrated history.

Stegner, Wallace. *Beyond the Hundredth Meridian: John Wesley Powell and the Second Opening of the West*. Boston: Houghton Mifflin Co., 1954. Reprint. New York: Penguin Books, 1992. New introduction by Bernard De Voto. The standard biography. Stegner brings a novelist's gifts to his compelling narrative. Illustrations juxtapose early artists' renderings of the Grand Canyon with some of the first photographs of the region.

Wild, Peter. *Pioneer Conservationists of Western America*. Missoula, Mont.: Mountain Publishing Co., 1979. A brief, breezy, superficial account which contains a chapter on Powell's explorations and theories.

James E. Fickle

PYTHEAS

Born: c. 350-325 B.C.; Massalia, Gaul
Died: After 300 B.C.; perhaps Massalia, Gaul

Pytheas undertook the first lengthy voyage to the North Atlantic and may have circumnavigated England. This knowledge of the West, together with his astronomical observations, provided the basis for centuries of study.

Early Life

It is a special characteristic of the study of antiquity that the fewer facts scholars know about a figure, the more they seem to write about him. So it is that an enormous bibliography about Pytheas of Massalia, the first known man to explore the far reaches of the North Atlantic, has evolved.

The time period of Pytheas' voyage has been determined with some certainty. Pytheas seems to have used a reference work which dates to 347 B.C., but since he is not mentioned by Aristotle, perhaps the voyage had not occurred before Aristotle's death in 322 B.C. Also, according to Strabo, Pytheas is quoted by Dicaearchus, who died circa 285 B.C. Thus, the voyage most definitely occurred between 347 and 285 B.C. At this time Carthage was the leading city of the western Mediterranean and controlled all traffic in and out of the Pillars of Hercules (Gibraltar). It is, therefore, sometimes claimed that Pytheas could have escaped this blockade only while Carthage was distracted in the war with Syracuse. If these assumptions are correct, the voyage took place between 310 and 306 B.C. Further, since Pytheas was surely a mature adult when he undertook the journey, scholars place his birth roughly between 350 and 325 B.C.

The date for the voyage is important, for it is believed that Pytheas opened the world of the West to Greek exploration at the same time that the wonders of the Far East were trickling back to the Mediterranean as a result of the conquests of Alexander the Great. The cosmopolitan Hellenistic age was being born and a quest for knowledge of far-off lands and their marvels was to play a large role in it. Apart from this tenuous but probable date, only two firm facts about Pytheas' life—his financial condition and his place of origin—are known. Polybius, also quoted by Strabo, sneers at Pytheas' voyage, asking if it was likely that a private citizen, and a poor one at that, ever undertook such a venture. Although Polybius was far from impartial, this comment may indicate that the voyage was state-sponsored.

Ancient authors are unanimous in calling him "Pytheas of Massalia," modern Marseilles. Modern texts often call him "Pytheas of Massilia," using the less accurate Roman form of the name. This place of origin is not unexpected, for Massalia, founded circa 600 B.C. by Phocaea in Asia Minor, was one of the most ambitious seafaring Greek towns. It soon controlled the coast, from its fine harbor down to modern Ampurias, seventy-five miles northeast of Barcelona. A Massaliote named Euthymenes was said to have sailed south along Africa until he saw a river filled with crocodiles (possibly the Senegal), and Massalia had early trading connections with metal-rich Tartessus in Spain. Friction with Carthage was inevitable as the two powers sought control of these rich trade routes. Into this tradition of Massaliote adventurism Pytheas was born, poor but ambitious.

Life's Work

Not a word of Pytheas' works remains. It has even been suggested that Pytheas' own works were not available to such authors as Diodorus Siculus (who wrote under Julius Caesar and Augustus), Strabo (who wrote under Augustus), and Pliny the Elder, who preserved for posterity meager fragments of Pytheas' research by quoting from or citing his works. Very often the information is secondhand, preceded by such phrases as "Polybius says that Pytheas claims that...."

Nevertheless, it is clear that Pytheas was remembered fondly as an astronomical scientist. Using only a sundial, he calculated the latitude of Massalia with remarkable accuracy. He noted first that the pole star was not really at the pole and was also the first to notice a relationship between the moon and the tides. Much of the information on latitudes and geography that he brought back from his voyages was deemed sufficiently accurate to be used by such famous ancient scholars as Timaeus, Hipparchus, and Ptolemy.

Pytheas the explorer, however, had another reputation entirely, neatly summed up by Strabo's calling him "the greatest liar among mortals." The nature and name of the work which reaped such abuse are unknown. The work may have been called "On the Ocean," "The Periplus" (meaning "voyage"), or "Travels Around the World." Modern scholars generally believe that it was a single work and that it recounted Pytheas' voyage. There is much to be said, however, for the theory that it was a general work of geography in which he reported his own firsthand observations, along with the rumors and reports he heard from others. If this is so, the scorn of later antiquity, relying on a spurious text, is more understandable. One can imagine the same comments being directed at Herodotus if only the more marvelous passages of his work had survived in this fashion.

With all that as warning, it is still customary to take the scattered references to Pytheas' voyage and reconstruct his route. If this approach is valid, his travels are impressive indeed. He left the Pillars of Hercules and cruised around Spain and the coast of France to the coast of Brittany and Ushant Island. Instead of continuing his coastal route as was customary for ancient mariners, he apparently struck out across the channel to Land's End, at the southwest tip of Britain at Cornwall. Here he described local tin mining. It is often asserted that Pytheas then circumnavigated the entire island of Britain. This belief is based on the fact that he describes the shape of the island correctly, describes its relationship to the coast better than did his critic Strabo, and, although doubling their true lengths, still correctly determines the proportion of the three sides. He probably made frequent observations of native behavior, and he may have conducted investigations inland. Diodorus, probably relying on Pytheas, reported correctly that the natives' huts were primitive, that they were basically peaceful but knew the chariot used for warfare, that they threshed their grain indoors because of the wet climate, and that they brewed and consumed mead.

Pytheas undoubtedly passed by Ireland, although no specific mention of this is found. It is often claimed, however, that his observations on the island enabled subsequent ancient geographers to locate it accurately on their maps. He apparently moved on to the northern tip of Britain, where he blandly described incredible tides eighty cubits (120 feet) high. Modern scholars see in this the gale-enhanced tides of the Pentland Firth.

It is the next stop on Pytheas' voyage which causes the greatest discussion. Pytheas claims that the island of Thule lay six days to the north of Britain and only one day from the frozen sea, sometimes called the Cronian Sea. Here, he states, days have up to twenty hours of sunlight in summer and twenty hours of darkness in winter. As if that information were not sufficiently incredible, he claims that the island lay in semicongealed waters in an area where earth, sea, and air are all mixed, suspended in a mixture resembling "sea lung" (perhaps a sort of jellyfish).

Where, if anywhere, is this Thule? Pytheas only claims that he saw the sea lung, getting the rest secondhand. Some parts of his tale ring true, such as long northerly days of light or darkness and a mixture of fog, mist, and slush so thick that one cannot tell where sea ends and sky begins. Scholars variously identify Thule as Iceland, Norway, the Shetland Islands, or the Orkney Islands, but no one solution is entirely satisfactory.

Pytheas soon turned south and completed his circumnavigation until he recrossed the channel. Here, again, there are problems, for he claims to have visited amber-rich lands as far as the Tanais River, acknowledged as the boundary between Europe and Asia. Scholars claim either that Pytheas reached the Vistula River and thus, remarkably, the heart of the Baltic Sea or that he stopped at the Elbe River. In either case, it is generally assumed that from there he retraced his steps along the European coast and returned home. Even by the most conservative estimates, he had traveled a minimum of seventy-five hundred miles in ships designed for the Mediterranean and manned by sailors unfamiliar with the rigors of the northern seas.

Impact

How can one assess a man and voyage so beset with problems of historicity? Did Pytheas in fact make a voyage at all? Was it a single voyage or were there two—one to Britain and one to the land of amber? In either case, how far did he go and how much information is from his own experience and how much is from what he learned through inquiry?

Barring the remarkable discovery of a long-lost Pytheas manuscript, these questions will never be answered. A coin from Cyrene, found on the northern coast of Brittany and dating to this time, has been cautiously set forth as evidence of Greek intrusion at this date, but the caution is well deserved.

Yet, despite the poor evidence and the hostility of the ancient authors, scholars can gauge Pytheas' importance from the impact he had on those who came after him. Pytheas opened Greek eyes to the wonders of the West, and it was his reports, for better or worse, which formed the basis for all writers on this area of the world for two centuries to come. In the same way, his scientific observations were respected and used by the best geographical minds of antiquity.

Still, it is highly likely that Pytheas did undertake a voyage himself and that he pushed fairly far to the north. Several thorny problems are solved if one believes that many of his wilder statements were not based on firsthand information but on tales he heard along the way. Much of the difficulty regarding Thule, for example, disappears when one views Pytheas' "discoveries" in this light.

The purpose of this voyage is also unclear. Some have hailed it as the first purely scientific voyage known to humankind. Yet if Pytheas was in fact a poor man and thus had public funds behind him, it is highly unlikely that the elders of Massalia would have found reports of sea lung proper repayment for their investment. It is wiser to see the voyage as primarily commercial, aimed at rivaling Carthaginian trade routes to lands rich in tin and amber, although Pytheas clearly lost no opportunity to engage in scientific enquiry along the way. (To be sure, his entire trip north of Cornwall seems guided more by a sense of adventure than of mercantilism.)

The world soon forgot about Pytheas' contribution to Massaliote trade routes. In fact, there is no evidence that an increase in trade followed his maiden voyage. Less ephemeral were Pytheas' tales of gigantic tides, sea lung, or Thule. His appeal extends into modern times, as the term "ultima Thule" remains a synonym for "the ends of the earth."

Bibliography

Bunbury, E. H. *A History of Ancient Geography Among the Greeks and Romans from the Earliest Ages Till the Fall of the Roman Empire.* 2d ed. Mineola, N.Y.: Dover Publications, 1959. A very sensible and cautious reconstruction of the probable circumstances surrounding Pytheas' voyage.

Carpenter, Rhys. *Beyond the Pillars of Heracles.* New York: Delacorte Press, 1966. A very lengthy section devoted to Pytheas treats several issues in great detail. An unorthodox date for Pytheas' life is to be rejected but the discussion of Thule is very well done.

Cary, Max, and E. H. Warmington. *The Ancient Explorers.* London: Methuen and Co., 1929. Rev. ed. Baltimore: Penguin Books, 1963. A somewhat uncritical re-creation of the voyage, with a tendency to gloss over several of the thornier questions.

Whitaker, Ian. "The Problem of Pytheas' Thule." *Classical Journal* 77 (1982): 148-164. A fine, careful study not only of Thule but also of most of the crucial problems surrounding Pytheas. Contains excellent documentation and bibliography, with translations of crucial passages from ancient authorities.

Kenneth F. Kitchell, Jr.

SIR WALTER RALEGH

Born: 1552 or 1554; Hayes Barton, Devon, England
Died: October 29, 1618; London, England

Ralegh's vision and enterprise paved the way for English settlement in North America.

Early Life

Walter Ralegh's birth date is even more uncertain than that of his contemporary William Shakespeare, but the dates of their deaths are precisely recorded, because by then they were among the most famous men of their time. Similarly, their family names are spelled in various ways. More than seventy spellings are recorded for Ralegh, the form he preferred in the second half of his life.

Ralegh is often designated as having been born in 1552, though 1554 accords with depositions he made in lawsuits. In any case, his birth occurred in the farm, or Barton, of Hayes, near East Budleigh on the south coast of Devon. His father was a gentleman farmer, who, like some of his relatives and other adventurous men of southwestern England, made money from maritime ventures, including privateering. Young Walter assuredly learned much about seafaring, as imaginatively depicted in Sir John Everett Millais' famous painting of Walter and another boy sitting on the beach, listening enthralled to a sailor's tale. Famous as he was to become by seafaring, however, Walter first made his mark as a soldier on land. At the end of the 1560's, he was campaigning in France as one of the volunteers fighting for the Protestant Huguenots against the Catholics, an experience which helped to shape his anti-Catholic attitude for the rest of his life. By 1572, he was an undergraduate at Oriel College in Oxford University, but within two years he left without taking a degree, a common practice then. In 1575, he enrolled in the Middle Temple, one of the Inns of Court in London, though he did not complete his legal education. No doubt he acquired knowledge of city and court ways.

Life's Work

In 1578, Ralegh sailed from Plymouth in Devon as captain of one of the ships under the command of his half brother, Sir Humphrey Gilbert, who held the charter to settle new lands for the Crown. The expedition aimed to explore and colonize the coast of North America. Bad weather drove the other ships back to England, but Ralegh persevered and reached the Cape Verde Islands, four hundred miles west of Africa.

After obtaining a minor post at court, in 1580 he was given command of a company of soldiers sent to help suppress rebellion in Ireland. He was involved in savage fighting, he befriended the poet Edmund Spenser, and he got Alice Gould pregnant. (He provided in his will for their illegitimate daughter and found Alice a well-to-do husband.) According to one account, "Ralegh coming out of Ireland to the English Court in good habit (his clothes then being a considerable part of his estate) found the Queen walking [in] a plashy place." He immediately "spread his new plush cloak on the ground, whereon the Queen trod gently, rewarding him afterward." This story, reported some eighty years later in Thomas Fuller's *The History of the Worthies of England* (1662), may be apocryphal, but it contains two indisputable truths: paintings and miniatures of Ralegh show that he dressed in the most opulent styles of the period, and he quickly became one of Queen Elizabeth I's favorite courtiers. In 1583, she gave him Durham House, a mansion on the north bank of the Thames, east of Westminster Abbey, and in 1584 the profitable monopolies of "the farm of wines" (by which he was authorized to charge every vintner in the realm one pound a year to sell wine) and the license to export woolen broadcloths. Also in 1584, he became a member of parliament for Devon and soon afterward Vice Admiral of Devon and Cornwall, Lord Lieutenant of Cornwall, and Lord Warden of the Stannaries (the tin mines of Cornwall). In January, 1585, the queen bestowed a knighthood on him and later made him captain of her guard.

Ralegh was adept at flattering the queen in Petrarchan poems

Archive Photos

praising her beauty, power, and influence. More tangibly, he would present to her the new lands of "Virginia," now North Carolina, where the expedition he equipped had landed in 1584. The following year, he sent about one hundred men to Roanoke Island on its coast, but they returned after the hardships of the first winter proved too severe for them. In 1587, a third expedition brought more than one hundred men and women to the site. The first child was born to the colonists on August 18 and christened Virginia Dare. Dealing with the Spanish Armada prevented a relief expedition from coming out until 1590, by which time the colonists had vanished. Although this lost colony was Ralegh's last colonizing attempt in North America, his efforts paved the way for the establishment there of an English-speaking empire in the early seventeenth century and prevented the northward spread of the Spanish Empire.

By the end of the 1580's, Ralegh could have been satisfied both by his personal advancement and by England's success at Spain's expense. Yet his fortunes were being undermined. The dashing and ambitious Robert Devereux, Earl of Essex, arrived at court and soon became the aging queen's latest favorite. Meanwhile, Ralegh became involved with one of her ladies-in-waiting, Elizabeth Throckmorton, who was a dozen years younger than he and some thirty years younger than the queen. By November, 1591, if not earlier, Ralegh was secretly married to Throckmorton. A son was born in March, 1592, but seems not to have survived for long. A second son, Walter, was born in 1593 and a third, Carew, in 1604.

In 1592, Queen Elizabeth had put Ralegh in command of an expedition against Panama, though forbidding him to sail beyond Spain. While he was at sea, she learned of his secret marriage, and on his return she had him and his wife imprisoned separately in the Tower of London. When the expedition returned with a captured Portuguese galleon laden with riches from the East Indies, Ralegh was sent to Dartmouth to make sure that the queen's share of the booty—and his share, which she appropriated—were not looted. In effect, he was obliged to buy his pardon.

Released from prison but banished from court, Ralegh and his wife withdrew to the Dorset estate of Sherborne, which he had begged from the queen while still in favor and which he now set about rebuilding. This activity, however, did not satisfy his ambitions. After Sir Humphrey Gilbert drowned on the voyage home from Newfoundland in 1583, Ralegh had acquired Gilbert's charter to explore and settle new lands. He now focused on South America, source of the wealth carried to Spain in the ships which Sir Francis

DRAKE, Ralegh, and others captured at sea. In one such action Sir Richard Grenville lost his life in a heroic (though perhaps ill-judged) rearguard action at the Azores in 1591; Ralegh glorified this event in *A Report of the Truth of the Fight About the Isles of Açores This Last Summer* (1591; also known as *The Last Fight of the Revenge*), his first published book. (Individual poems by or attributed to him were published from 1576 onward, but he never published a collection of his poems.) Ralegh was convinced that in the hinterlands of Guiana, now Venezuela, lay the fabulously rich empire of El Dorado. In 1595, he led an expedition to Guiana and on his return promptly wrote and published *The Discovery of the Large, Rich, and Beautiful Empire of Guiana* ... (1596), in which he argued that abundant gold could be found there, and that friendly Indians were eager to overthrow their cruel Spanish oppressors and welcome the benign rule of Elizabeth.

Whatever the queen thought of this argument, Ralegh was soon employed on a different venture. He was given command of a squadron in the 1596 expedition against Cadiz, under the leadership of Lord Admiral Howard and Essex. Ralegh boldly led his ships against the harbor defenses and suffered a leg wound which left him using a cane for the rest of his life. His spirited initiative was not shared by the commanders, whose temporizing failed to secure the fullest spoils possible.

The following year, he and Essex led another expedition to seize a Spanish treasure fleet off the Azores. Again, Essex's inadequacies and Ralegh's courage were revealed, the only gain being the temporary capture of the port of Fayal, which Ralegh achieved by leading his men ashore under fire. These events did nothing to assuage the rivalry between the two courtiers. When the irrationally ambitious Essex raised his abortive rebellion against the queen in 1601 and was executed for doing so, suspicion that Ralegh had contributed to his doom was widespread.

While Essex was ruining himself, Ralegh was improving the trade and fortifications of the isle of Jersey, of which the queen made him governor in 1600. This was to be his last advancement, however, because with her death in 1603 his fortunes plummeted. The new sovereign, James I, was strongly biased against Ralegh, reportedly greeting him with the words, "I have heard rawly of thee" and soon depriving him of his positions. Rumored to be discontented, as he might well have been, Ralegh was suspected of treasonous conspiracies against the new king from Scotland. In 1593, he had been exonerated when tried for atheism for his association with the "School of Night," a group of skeptics and freethinkers which included Christopher Marlowe. In 1603, the charges were to be even more implausible but more far-reaching. Ralegh was accused of being in Spanish pay to seek a new policy of peace toward Spain and to be part of a conspiracy to depose James and replace him with his cousin Arabella Stuart. Though Ralegh's position on these matters is not entirely clear, his trial was conducted with appalling injustice and venom, and in spite of the splendid speeches he made in his defense, a rigged jury guaranteed that he would be found guilty and sentenced to death.

Perhaps because executing the "last Elizabethan" hero was deemed to be impolitic, Ralegh was not put to death but imprisoned in the Tower of London again, this time for almost thirteen years. Again refusing simply to languish in royal disfavor, Ralegh wrote letters containing exaggerated professions of regard for James and humiliating pleas for pardon. He had hundreds of books brought in and embarked on writing *The History of the World* (1614). This monumental undertaking went as far as 133 B.C., and although Ralegh does not refer to the sovereigns he served, James denounced the book as "too saucy in censuring princes" and tried to suppress its publication. James's enmity was not shared by his queen, Anne of Denmark, and their son Prince Henry, both of whom often visited Ralegh in the Tower. Ralegh served as tutor to Prince Henry, for whom he wrote *The History of the World* and whose premature death in 1612 at the age of eighteen caused Ralegh to stop work on the book. The death was a double blow to Ralegh, not only because of the prince's announcement that "No one but my father would keep such a bird in a cage," but also because the manly and chivalric prince seemed likely to be the inspiring monarch that James was not.

James's attempts to secure a substantial dowry from a proposed marriage between his younger son Charles and a Spanish princess had been frustrated by 1616. Hearkening to Ralegh's continual claim that gold could be extracted from Guiana, James released Ralegh to organize and lead an expedition there. At the same time James secretly assured the Spaniards that if Ralegh came into conflict with them, his life would be forfeit. Ralegh was now in his sixties and had suffered several strokes. By the time the expedition neared Guiana in late 1617 he was so ill with fever that he had to delegate command of the party that went up the river Orinoco to his trusted second in command, Lawrence Keymis. At the fort of San Thomé, the party got into a fight with the Spaniards, during which Ralegh's son Walter was killed. No gold was found; after returning to the ships, Keymis committed suicide.

Ralegh returned to England a shattered man and was soon imprisoned yet again in the Tower. After having been condemned to death in 1603 on the charge of conspiring to make peace with Spain, he was now to be executed for making war with Spain. The sentence of fifteen years earlier was carried out on October 29, 1618. A huge crowd gathered in the Old Palace Yard at Westminster and Ralegh, elegantly dressed, delivered a speech of nearly an hour in which he defended himself against the charges brought against him and committed himself to the mercy of God. Declining a blindfold, he laid his head on the block and told the hesitant executioner, "What dost thou fear? Strike man, strike!" The headsman needed two blows to sever Ralegh's head, which was carried away by his widow, while his body was buried in the nearby church of St. Margaret's, Westminster.

Impact

Often disliked as a proud, ambitious upstart during his rise, Sir Walter Ralegh, by the courage and grace with which he faced his end, won widespread sympathy as a political martyr. Among those who witnessed his execution were some of the men who would lead the Great Rebellion of the 1640's against the autocratic despotism of the Stuart monarchy. Ironically, therefore, the beheaded victim of King James became an influence on those who would behead James's son King Charles I thirty years and three months later.

Ralegh's life has an aura of mystery to it. He had the characteristics of both a hero and a scoundrel. Further, his vision of the possibilities of empire for England in the Americas, although persuasively supported by his gift as a writer and his daring as an explorer and soldier, would not be realized in his lifetime. Still, he captured the imagination of the English people, and his enterprises and plans were brought to completion by others. Thus, Ralegh can be seen as a colorful, gifted person who failed to be the historical force he might have become.

Bibliography

Adamson, J. H., and H. F. Folland. *The Shepherd of the Ocean: An Account of Sir Walter Ralegh and His Times*. Boston: Gambit, 1969. Sets Ralegh's life in its historical and political contexts and devotes ample space to summarizing his literary work and relationships.

Armitage, Christopher M. *Sir Walter Ralegh: An Annotated Bibliography*. Chapel Hill: University of North Carolina Press, 1987. Contains nearly two thousand items by and about Ralegh, from 1576 to 1986.

Edwards, Philip, ed. *Last Voyages—Cavendish, Hudson, Ralegh: The Original Narratives*. New York: Oxford University Press, 1988. A selection of writings documenting the voyages of the three great English explorers.

Hammond, Peter. *Sir Walter Ralegh*. London: Pitkin Books, 1978. A concise biography, with abundant pictures of people and places of significance in Ralegh's life.

Jones, H.G., ed. *Ralegh and Quinn: The Explorer and His Boswell*. Chapel Hill: North Caroliniana Society, 1987. A wide-ranging set of papers from the 1987 International Conference on Ralegh, at which David Beers Quinn, emeritus professor at the University of Liverpool, was honored for his many publications in the field.

Lefranc, Pierre. *Sir Walter Ralegh, écrivain: L'Oeuvre et les idées*. Quebec: Presses de l'Université Laval, 1968. Considers Ralegh's mind and art, analyzes the evidence for his authorship of the poems and prose attributed to him, and evaluates his development as a writer.

Mills, Jerry Leath. *Sir Walter Ralegh*. Boston: G. K. Hall, 1986. A year-by-year listing, from 1901 to 1984, of books and articles about Ralegh, with often extensive annotation by the compiler.

Ralegh, Sir Walter. *The Discoverie of the Large, Rich, and Bewtiful Empyre of Guiana*. Transcribed, annotated, and introduced by Neil L. Whitehead. Manchester: Manchester University Press, 1997. Of particular interest in studying Ralegh's contribution to the exploration of the New World.

———. *The History of the World*. Edited by C. A. Patrides. London: Macmillan, 1971. The most substantial modern selection from this huge work, with an analysis of Ralegh's achievement as a writer of history.

———. *Selected Prose and Poetry*. Edited by Agnes M. C. Latham. London: University of London Athlone Press, 1965. A selection by the editor of the standard edition of *The Poems of Sir Walter Ralegh* (1929).

———. *Selected Writings*. Edited by Gerald Hammond. Harmondsworth: Penguin Books, 1986. A convenient modern selection of Ralegh's poems, prose works, and letters.

Wallace, Willard M. *Sir Walter Ralegh*. Princeton: Princeton University Press, 1959. Covers Ralegh's life and pays considerable attention to his literary work.

———. *The Works of Sir Walter Ralegh*. Oxford: Oxford University Press, 1829. 8 vols. Reprint. New York: Burt Franklin, 1965. Volume 1 contains the early biographies of Ralegh by W. Oldys and T. Birch; volumes 2-7 contain *The History of the World*; and volume 8 contains miscellaneous essays, poems, and letters, many now considered to have been foisted on Ralegh after his death.

Christopher Armitage

KNUD JOHAN VICTOR RASMUSSEN

Born: June 7, 1879; Jakobshavn, Greenland
Died: December 21, 1933; Gentofte, Denmark

A pioneer Arctic explorer, Rasmussen was best known for his seven Thule expeditions. In the fifth, the most famous of these, he crossed North America from Greenland to the Bering Strait. A celebrated ethnographer, Rasmussen studied the folkways of the Eskimos and published many works about the peoples and places of Arctic America.

Early Life

Knud Johan Victor Rasmussen was born on June 7, 1879, in the Lutheran parsonage at Jakobshavn, Greenland. This Danish settlement was situated halfway up the western coast of Greenland. The eldest son of Christian Rasmussen, a Danish missionary in Greenland for twenty-eight years, who later became a lector in Greenlandic studies at the University of Copenhagen, Knud was exposed to exploration and ethnography in early childhood. His father took as his parish the entire northern half of colonized Greenland, often working his way by dogsled up the west coast of the island to visit his five remote preaching stations. An excellent linguist who later produced both a Greenlandic grammar and dictionary, the elder Rasmussen taught Knud to regard all Greenlanders as his brothers, and Knud responded by learning their ways and developing a love for them that never waned.

Rasmussen's mother was herself part Eskimo. Her father, Knud Fleischer, had been born in Greenland of Norwegian parents. Becoming a colonial administrator for the Danes as well as a successful trader, Fleischer married an Eskimo woman. Young Rasmussen grew up celebrating his dual heritage—the Scandinavian (Danish and Norwegian) and the Eskimo.

Rasmussen recalled his childhood as a happy one. From the parsonage, he could view Disko Island, the largest off the coast of Greenland, as well as the great glacier and the spring icebergs. Fascinated by the North, Rasmussen rejoiced in a childhood trip with his father and Riis Carstensen, an explorer, to visit his uncle, Carl Fleischer, who headed the Danish settlement at Qeqertak. This Greenland childhood determined the direction of Rasmussen's later life. Two additional influences affected Rasmussen's development. In 1888, Fridtjof NANSEN attempted the first complete crossing of Greenland, an adventure that had a profound influence on the lad. The impact of his Aunt Helga, his first teacher, was equally decisive. It filled him with a profound love for the ways of Greenland.

Reluctantly, Rasmussen left Greenland for Denmark. Failing his entrance examinations for the Herlufsholm School, Rasmussen studied in Copenhagen. He was not a particularly good student. Completing his baccalaureate education at the University of Copenhagen (Rasmussen later was awarded a Ph.D. by his alma mater

Knud Johan Victor Rasmussen (third from left). Library of Congress

and an LL.B. by the University of Edinburgh), Rasmussen flirted with several occupations, such as acting, singing, and journalism. As a correspondent for the *Christian Daily* and the *Illustrated Times*, he went to Stockholm to cover the Nordic games; then, at age twenty-one, he went to Lapland to study reindeer breeding. Travels in Scandinavia's Northland, to Narvik and Tomso, reinforced his fascination with the Arctic.

Life's Work

At age twenty-three, Rasmussen began his life's work. He joined the Danish Literacy Expedition of Mylius-Erichsen, an ethnographer, Jorgen Brønland, a catechist, Count Harald Moltke, a painter and illustrator, and Alfred Bertelsen, a doctor, on an expedition to visit the most northern tribe in the world, the Polar Eskimos of upper Greenland. This voyage of 1902-1904 was followed in 1905 by an assignment from the Danish government to travel in Greenland with a group of Lapps to determine the feasibility of introducing the reindeer as an addition to the Eskimo economy. For the next two years, 1906-1908, Rasmussen lived among the Polar Eskimos, studying their folklore. By then it was becoming obvious that Rasmussen's ability to travel and hunt like the Eskimos was a phenomenal asset. He could speak their languages fluently and maintain friendly relations with them. Rasmussen was able to record much of their oral tradition before it disappeared with the onset of modern civilization.

Returning to Denmark in 1908, Rasmussen married Dagmar Andersen, daughter of Niels Andersen, state counselor, chairman of the Employers' Association, and considered one of Denmark's major entrepreneurs. Friends considered this marriage a major source of strength for Rasmussen. Within a year, Rasmussen had returned to the Arctic, serving the Danish government on an expedition for educational purposes in 1909. This fired his imagination and caused him to envision the possibility of founding a permanent base for additional explorations.

At the age of thirty, in 1910, Rasmussen established Thule, a center for trade and exploration among the Polar Eskimos. Trade in manufactured goods provided the economic support, but the real purpose of this base on the northwest coast of Greenland was not commercial but scientific. Thule became the starting point for seven expeditions. Rasmussen's timing was excellent. The discovery of the North Pole in 1909 had aroused considerable interest in the Arctic. Danish claims to the north of Greenland were being contested, and Rasmussen saw such a settlement as Thule as critical to establishing Danish sovereignty over the region. This opinion was vindicated in 1933, when the International Court of Justice at the Hague ruled against Norway and in favor of Denmark, recognizing Copenhagen's claims to all of Greenland. Following a lecture tour to raise funds for building Thule, Rasmussen sailed to the Arctic. The harbor at Thule was open only twenty-five days of the year (August 1-25), and the environment was harsh. Rasmussen coped with these conditions and became the first to cross Melville Bay by sledge, demonstrating the feasibility of exploration from Thule.

On April 8, 1912, together with explorer Peter Freuchen, a longtime friend, Rasmussen led the first Thule expedition, crossing the Greenland ice cap from Thule to Independence Fjord. This feat had been attempted only once before, by NANSEN in 1888, an event that had inspired Rasmussen as a child. This trip allowed Rasmussen to study Eskimo life and to formulate his theory as to their origins. Postulating their Asian origin, Rasmussen believed that American Indians and Eskimos were descended from prehistoric immigrants who came to the Americas across the Bering Strait. Upon the completion of the first Thule expedition, Rasmussen returned to Denmark to report on his scientific progress and to see his three-year-old daughter for the first time.

Though Denmark remained neutral during World War I, the European conflagration had consequences for the far North. Rasmussen continued his work, however, and a mapping expedition in 1914 was followed in 1916-1918 by a survey of the north coast of Greenland. In 1918, following a visit to Denmark, Rasmussen set out for Angmagssalik in eastern Greenland on an ethnographic expedition to collect Eskimo tales. This was completed in 1919. On the two-hundredth anniversary of the arrival in Greenland of Hans Egede, the pioneer Lutheran missionary, there was a royal visit by the King of Denmark to the island. This event in 1921 honoring "the

Apostle of Greenland" encouraged Rasmussen to think in terms of further discoveries.

The fifth Thule expedition, Rasmussen's most famous journey, lasted from 1921 to 1924, and he explored Greenland, Baffin Island, and the Arctic Coast of America, the longest dogsled journey in history. Rasmussen traversed the American Arctic from the Atlantic to the Pacific, conducting a scientific study of virtually every Eskimo tribe in that region. The expedition began on September 7, 1921, at Upernavik and went from Greenland to the Bering Strait, arriving at Point Barrow, Alaska, on May 23, 1924. During this trip, Rasmussen traced Eskimo migration routes and observed the essential unity of Eskimo culture.

Rasmussen was an excellent communicator, and his works were widely published in Danish, Greenlandic, and English translation. Rasmussen's works included travelogs, collections of Eskimo mythology and songs, and scientific texts, as well as writings of cartographic, ethnographical, and archaeological significance. *Under nordenvindens svobe* (1906) and *Nye mennesker* (1905) appeared in English translation in 1908 under the single title *The People of the Polar North: A Record* and established his reputation. *Grønland Langs Polhavet: Udforskningen af Grønland fra Melvillebugten til Kap Morris Jesup* (1919; *Greenland by the Polar Sea: The Story of the Thule Expedition from Melville Bay to Cape Morris Jessup*, 1921) introduced the earth's largest island to readers throughout the Western world and was followed within the decade by his account of the most extensive expedition yet to explore the Arctic, published as *Fra Grønland til Stillehavet* (1925; *Across Arctic America: Narrative of the Fifth Thule Expedition*, 1927). Rasmussen's work also included collections of Native American literature such as *Myster og sagn fra Grønland* (1921-1925; myths and sagas from Greenland).

Impact

Knud Johan Victor Rasmussen was honored by the world for his many scientific contributions and was a Knight of the Royal Order of Dannebrog (Denmark), a Commander of the Order of Saint Olav (Norway), a Commander of the White Rose (Finland), a Knight of the Royal Order of the North Star (Sweden), and a recipient of a Golden Medal of Merit from the Danish king, among other awards. Rasmussen was a member of many distinguished learned societies, including the Norwegian Geographical Society and the equivalent geographical societies of Sweden, Italy, and the United States as well as the Explorers' Club of New York and the Scientific Society in Lund, Sweden.

Explorer of the Arctic and famed ethnographer of the American Eskimos, Rasmussen was honored with doctorates from Danish and British universities and the Knud Rasmussen room in the National Museum in Copenhagen recalls his memory. More than sixteen thousand artifacts in that museum testify to the thoroughness of his work. On December 21, 1933, Rasmussen died near Copenhagen, at Gentofte, Denmark, of food poisoning contracted during his final expedition, complicated by influenza and pneumonia.

Bibliography

Croft, Andrew. *Polar Exploration.* 2d ed. London: Adam and Charles Black, 1947. More than a general survey of polar expeditions, this volume focuses on the more prominent explorations of the Arctic regions in the twentieth century. With eight maps and twenty-two illustrations, this text is organized into two parts. Part 1, entitled "The Arctic Regions," is especially relevant to the life of Rasmussen; it surveys the scientific exploration of the North and contains valuable discussion of Rasmussen's contribution to geographical knowledge in Greenland and Canada.

Freuchen, Peter. *Arctic Adventure: My Life in the Frozen North.* New York: Farrar and Rinehart, 1935. Freuchen was Rasmussen's best friend, and together they shared many interests and experiences. Enhanced with illustrations and maps, this book is more than a recollection of one man's life in the Arctic. Contains interesting vignettes of the region, its conditions, and peoples. Invaluable personal recollections and anecdotes.

_____. *I Sailed with Rasmussen.* New York: Julian Messner, 1958. This work is not an exhaustive scholarly work on Rasmussen but rather a collection of impressions of a dear friend. A vivid description that is supplemented by useful illustrations.

Stefansson, Vilhjalmur. *Greenland.* Garden City, N.Y.: Doubleday, Doran, 1942. An older work, this history of Greenland from the earliest times until the start of the 1940's remains a valuable introduction to the world that Rasmussen knew and loved. Readable and reliable, Stefansson's survey conveys a feel for a region that is as large as the combined twenty-six states east of the Mississippi. Particularly helpful are references to Rasmussen's works.

Williamson, Geoffrey. *Changing Greenland.* Introduction by Ole Bjørn Kraft. New York: Library Publishers, 1954. This survey of the history of Greenland from the arrival of the Vikings to the major changes of the 1950's is organized into two main sections. Part 1, entitled "Old Orders," helps place the life and labors of Rasmussen in proper chronological and sociological context.

C. George Fry

SALLY RIDE

Born: May 26, 1951; Encino, California

An astronaut for the National Aeronautics and Space Administration (NASA) and the first American woman to fly in space.

Early Life

Sally Kristen Ride was born on May 26, 1951, in Encino, a suburb of Los Angeles, California. She was the older of two daughters born to Dale B. Ride, a member of the faculty at Santa Monica Community College, and his wife, Joyce. Sally's parents were active as elders in their Presbyterian church, and Joyce Ride often volunteered her time as an English tutor to foreign-born students and as a counselor at a women's prison.

Sally's parents encouraged her competitive spirit in academics and in athletics. Sally was a born athlete and often played the rough and tumble games of football and baseball with the neighborhood boys. Ride began playing tennis, a less hazardous sport, at the request of her mother. Under the tutelage of tennis great Alice Marble, Sally quickly excelled in this sport and became proficient enough to rank eighteenth nationally. Her excellence in tennis earned her a partial scholarship to Westlake School for Girls, a private preparatory school in Los Angeles. At the preparatory school, Sally became interested in the study of physics through the influence of her science teacher, Elizabeth Mommaerts, and, for the next five years, science and tennis competed for Sally's time and attention.

In 1968, Sally enrolled at Swarthmore College in Pennsylvania as a physics major, but left after three terms to concentrate on her tennis game after winning a national collegiate tennis tournament. Although she was a top-ranked college player, she realized that she did not have the talent to advance to professional tennis. Sally returned to college in 1970 and completed a double major in English

Express Newspapers/Archive Photos

literature and physics at Stanford University in California in 1973. After graduation, she briefly considered continuing with Shakespeare in graduate school, but settled on astrophysics to further her dream of working for NASA.

Life's Work

Sally Ride began her path to fame while completing work on her doctoral dissertation at Stanford. One day she read an announcement in the campus newspaper indicating that NASA was seeking young scientists to serve as "mission specialists." Acting on impulse, she applied to join the astronaut program, which had lifted its long-standing ban against women in order to attract additional qualified scientists willing to forgo high salaries in order to work on the new space shuttle program. To Ride's surprise, she made it through the preliminary screening process to become one of the finalists. In 1977, she was flown to the Lyndon B. Johnson Space Center outside Houston, Texas, for exhausting interviews and fitness and psychiatric evaluation tests. After three months of rigorous testing, Sally Ride officially became an astronaut. In 1978, shortly after earning her Ph.D. degree, she reported to the Johnson Space Center to begin the intensive training required of NASA mission specialists.

In the first year of training, Ride learned parachute jumping and water survival techniques, the latter for the possibility that the shuttle might be ditched in the ocean. She also became acclimated to increased gravity forces, the force encountered during acceleration and deceleration back to earth, as well as to weightlessness. Ride took courses in radio communication and navigation and learned to fly a jet. Piloting a jet proved to be an enjoyable experience for Sally Ride, and she eventually acquired a pilot's license.

Throughout Ride's entire preparation time, NASA maintained its bureaucratic composure regarding its inclusion of women in the space shuttle program. There was no flamboyant talk about one giant step for womankind. Indeed, team player that she was, Ride insisted that her participation in the flight was "no big deal." Whether she liked it or not, news of her flight brought her instant celebrity. Newspapers and television reporters interviewed her again and again, and even President Ronald Reagan gave her an extra share of attention at a White House luncheon. Composer Casse Culver wrote and recorded a song entitled "Ride, Sally, Ride" to celebrate the event, and T-shirts urged the same.

Sally Ride was specifically requested by Navy Captain Robert L. Crippen, a veteran astronaut who had piloted the first shuttle mission in 1981. Crippen said, of his choice of Ride: "She is flying with us because she is the very best person for the job. There is no man I would rather have in her place." Ride, in her unassuming manner, simply stated that she had not become an astronaut to become "a historic figure," and that she believed it was "time that people realized that women in this country can do any job that they want to do."

Aboard the *Challenger*, Ride had duties in addition to her scientific work. She was chosen to sit behind mission commander Crippen and copilot Frederick Hauck to act as flight engineer during takeoff and landing. During the ninety-six orbits, she and her fellow mission specialist, John Fabian, worked in weightless conditions with the complex Canadarm—a fifty-foot remote "arm" used to move payloads in and out of the shuttle cargo bay. Ride and Fabian trained for two years on the ground with the computerized arm and became experts in its operation. Another task on the mission was to place Anik-C, a Canadian domestic communications satellite, in a geosynchronous orbit hovering above the equator. This satellite was designed to handle thirty-two color television channels. A second communications satellite, named Palapa-B and owned by Indonesia, was launched into orbit to carry voice, video, and telephone signals to southeast Asia.

Forty other experiments were conducted by the *Challenger* crew. These included studies of metal alloy manufacture, solar cell testing, growth of semiconductor crystals, and glass production. One experiment, devised by high-school students, was a project sending 150 carpenter ants into orbit in the shuttle cargo bay to see how weightlessness affects their social structure. The California Institute of Technology sent an experiment in which radish seedlings were subjected to simulated gravity to find the right gravitational force for best growth. Purdue University's experiment investigated how sunflower seeds germinated in zero gravity. The highlight of the mission was the deployment of a huge free-floating satellite in order to document its position with the first in-space color photographs before recapturing it. The satellite was then released again and snared once more. The crew repeated this procedure for nine and one-half hours before Ride captured the satellite for the last time and stowed it in the cargo bay for the trip home.

Sally's ride (a pun often used by the media) was only one sign of a major change in what could no longer be called the United States manned space program. Much of the daredevil aspect had gone out of space travel. The object was not simply getting into orbit but working there. In fact, Sally Ride recommended that space be used to study the planet Earth to NASA administrator James Fletcher, who made her his special assistant for Long Range and Strategic Planning.

Being a space pioneer was more important to Sally Ride than achieving celebrity as a woman astronaut. Specialists have since been recruited from the ranks of male and female scientists. For all the merits of the scientific and experimental aspects of the *Challenger* voyage, it was Sally Ride who provoked the world's curiosity. Cool, calm, and apparently controlled in any circumstance, Sally Kristen Ride hurtled through space aboard the 100-ton white and blue shuttle *Challenger*. Sally Ride showed that she was certainly made of "the right stuff."

Impact

Although two Soviet women preceded Sally Ride into space, they hardly left their mark on it. About 1963, in the early days of manned spaceflight, a twenty-six-year-old textile mill worker and amateur sky diver named Valentina Tereshkova was put on a rocket by the Soviet Union as a propaganda coup. Reports of that flight say that Tereshkova was sick for most of the three-day flight.

In August of 1982, the Soviets launched the second woman cosmonaut—a thirty-four-year-old test pilot named Svetlana Savitskaya. Her presence, however, was taken lightly by her colleagues.

Ride's mission came to signify the ascendancy of the mission specialist over the pilot. The close-knit brotherhood of test and fighter pilots who made up the original astronaut corps was diluted by those having a new kind of "right stuff"—the ability to do quadratic equations and conduct scientific experiments instead of mere fancy flying. Under these new guidelines, Ride was an ideal candidate not only because of her excellent scientific background but also because she exhibited the ability to learn new skills and solve problems readily. Sally Ride's experiences on the space shuttle earned for her the trust and high regard of her colleagues as well as the admiration of an entire nation.

Bibliography

Begley, Sharon. "Challenger: Ride, Sally Ride." *Newsweek* 101 (June 13, 1983): 20-21. An overview of Sally Ride's life and the results of her crucial decision to join the elite NASA astronaut group in preparation for missions on the space shuttle.

Fox, Mary Virginia. *Women Astronauts Aboard the Shuttle*. Rev. ed. New York: Julian Messner, 1987. Chronicles the experiences of the women who have been selected to participate in the space shuttle program. Aimed at young readers, this work focuses particular attention on Ride's experiences as the first American woman to fly in space while providing equally useful profiles of the various women who followed. Touches on Ride's decision to leave the astronaut program in 1987.

Golden, Frederic. "Sally's Joy Ride into the Sky." *Time* 121 (June 27, 1983): 56-58. In the magazine's "Space" section, Golden tells of Ride's experiences on the second orbiting flight of the space shuttle *Challenger* and includes some of Ride's own observations regarding the public's reaction to her flight.

Otto, Dixon P. *On Orbit: Bringing on the Space Shuttle*. Athens, Ohio: Main Stage Publications, 1986. Aimed at a general audience, this work examines the space shuttle from its design origins through its first twenty-five flights. Contains information about the crew members, payloads, and objectives of each flight as well as many black-and-white illustrations highlighting these missions. Provides a context for understanding Ride's experiences in the space shuttle program.

Ride, Sally, with Susan Okie. *To Space and Back*. New York: Lothrop, Lee & Shepard Books, 1986. Written for a young audience, this book describes the human side of being a member of an astronaut crew. Ride shares her personal experience of space travel on the space shuttle. The book does a fine job of revealing both the remarkable talents and the more ordinary characteristics of those individuals who have chosen to become space pioneers.

Ride, Sally, and Tam O'Shaughnessy. *Voyager: An Adventure to the Edge of the Solar System*. New York: Crown, 1992. Although this work is not specifically related to her shuttle experiences, Ride does draw upon her astrophysics background in order to create this popular account of the two Voyager spacecraft that were launched during the late 1970's in order to explore and transmit images of four of the solar system's most distant planets: Jupiter, Saturn, Uranus, and Neptune.

Jane A. Slezak

SACAGAWEA

Born: c. 1788; central Idaho
Died: December 20, 1812; Fort Manuel, Dakota Territory

The only woman who accompanied the Lewis and Clark Expedition in exploring much of the territory acquired through the Louisiana Purchase, Sacagawea assisted as guide and interpreter.

Early Life

Sacagawea (also Sagagawea, Sakakawea) was born into a band of Northern Shoshone Indians, whose base was the Lemhi Valley of central Idaho. Her name translates as "Bird Woman" (Hidatsa) or "Boat Pusher" (Shoshonean). The Northern Shoshone, sometimes referred to as Snake Indians (a name given them by the French because of the use of painted snakes on sticks to frighten their enemies), were a wandering people, living by hunting, gathering, and fishing. As a child, Sacagawea traveled through the mountains and valleys of Idaho, northwest Wyoming, and western Montana. In 1800, at about age twelve, Sacagawea and her kin were encamped during a hunting foray at the Three Forks of the Missouri (between modern Butte and Bozeman, Montana) when they were attacked by a war party of Hidatsas (also called Minnetarees), a Siouan tribe; about ten Shoshone were killed and Sacagawea and several other children were made captives. Sacagawea was taken to reside with the Hidatsas at the village of Metaharta near the junction of the Knife and Missouri Rivers (in modern North Dakota).

Shortly after her capture, Sacagawea was sold as a wife to fur trader Toussaint Charbonneau. A French-Canadian who had developed skills as an interpreter, Charbonneau had been living with the Hidatsas for five years. At the time that Sacagawea became his squaw, Charbonneau had one or two other Indian wives.

All that is known of Sacagawea for certain is found in the

journals and letters of Meriwether Lewis, William Clark, and several other participants in the expedition of the Corps of Discovery, 1804-1806, along with meager references in other sources. The LEWIS AND CLARK party, commissioned by President Thomas Jefferson to find a route to the Pacific and to make scientific observations along the way, traveled on the first leg of their journey up the Missouri River to the mouth of the Knife River, near which they established Fort Mandan (near modern Bismarck, North Dakota) as their winter headquarters. The site was in the vicinity of Mandan and Hidasta villages. Here the expedition's leaders made preparations for the next leg of their journey and collected information on the Indians and topography of the far West.

Life's Work

Sacagawea's association with the Lewis and Clark Expedition began on November 4, 1804, when she accompanied her husband to Fort Mandan. She presented the officers with four buffalo robes. Charbonneau was willing to serve as interpreter, but only on condition that Sacagawea be permitted to go along on the journey. After agreeing to those terms, Lewis and Clark hired Charbonneau. At Fort Mandan on February 11, 1805, Sacagawea gave birth to Jean-Baptiste Charbonneau. Thus, along with the some thirty men, the "squaw woman" and baby became members of the exploring group.

The Lewis and Clark Expedition set out from Fort Mandan on April 7, 1805. Charbonneau and Sacagawea at different times were referred to in the journals as "interpreter and interpretess." Sacagawea's knowledge of Hidatsa and Shoshonean proved of great aid in communicating with the two tribes with which the expedition primarily had contact. Later, when the expedition made contact with Pacific Coast Indians, Sacagawea managed to assist in communicating with those peoples even though she did not speak their language. Her services as a guide were helpful only when the expedition sought out Shoshone Indians in the region of the Continental Divide in order to find direction and assistance in leaving the mountains westward. Carrying her baby on her back in cord netting, Sacagawea stayed with one or several of the main groups of explorers, never venturing out scouting on her own. Little Baptiste enli-

Library of Congress

vened the camp circles, and Clark, unlike Lewis, became very fond of both baby and mother.

Several times on the westward journey Sacagawea was seriously ill, and once she and Baptiste were nearly swept away in a flash flood. In May of 1805, Sacagawea demonstrated her resourcefulness by retrieving many valuable articles that had washed out of a canoe during a rainstorm. LEWIS AND CLARK named a stream "Sâh-câ-ger we-âh (*Sah ca gah we a*) or bird woman's River," which at a later time was renamed Crooked Creek. Not the least of Sacagawea's contributions was finding sustenance in the forests, identifying flora that Indians considered edible. She helped to gather berries, wild onions, beans, artichokes, and roots. She cooked and mended clothes.

Reaching the Three Forks of the Missouri, Sacagawea recognized landmarks and rightly conjectured where the Shoshone might be during the hunting season. A band of these Indians was found along the Lemhi River. Sacagawea began "to dance and show every mark of the most extravagant joy ... sucking her fingers at the same time to indicate that they were of her native tribe." The tribe's leader, Cameahwait, turned out to be Sacagawea's brother (or possibly cousin). Lewis and Clark established a cordial relationship with Sacagawea's kinsmen, and were able to obtain twenty-nine horses and an Indian guide through the rest of the mountains. Coming down from the mountains, the exploring party made dugout canoes at the forks of the Clearwater River, and then followed an all-water route along that stream, the Snake River, and the Columbia River to the Pacific Coast. At the mouth of the Columbia River, just below present Astoria, Oregon, the adventurers built Fort Clatsop, where they spent the winter. Sacagawea was an important asset as the expedition covered the final phase of the journey. "The wife of Shabono our interpreter," wrote William Clark on October 13, 1805, "reconsiles all the Indians, as to our friendly intentions a woman with a party of men is a token of peace."

Besides her recognition of topography that aided in finding the Shoshones, Sacagawea's other contribution as guide occurred on the return trip. During the crossing of the eastern Rockies by Clark's party (Lewis took a more northerly route), Sacagawea showed the way from Three Forks through the mountains by way of the Bozeman Pass to the Yellowstone River. Lewis and Clark reunited near the junction of the Missouri and the Yellowstone. Sacagawea, Charbonneau, and infant Jean-Baptiste accompanied the expedition down the Missouri River only as far as the Hidatsa villages at the mouth of the Knife River. On April 17, 1806, they "took leave" of the exploring group. Clark offered to take Sacagawea's baby, whom Clark called "Pomp," with him to St. Louis to be reared and educated as his adopted son. Sacagawea, who consented to the proposal, insisted that the infant, then nineteen months old, be weaned first.

With the conclusion of the Lewis and Clark Expedition, details about Sacagawea's life become very sketchy. In the fall of 1809, the Charbonneau family visited St. Louis. Charbonneau purchased a small farm on the Missouri River just north of St. Louis from Clark, who had been named Indian superintendent for the Louisiana Territory. In 1811, Charbonneau sold back the tract to Clark. Sacagawea yearned to return to her homeland. Charbonneau enlisted in a fur trading expedition conducted by Manuel Lisa. In April of 1811, Sacagawea and Charbonneau headed up river in one of Lisa's boats. One observer on board at the time commented that Sacagawea appeared sickly.

Sacagawea left Jean-Baptiste Charbonneau with Clark in St. Louis. On August 11, 1813, an orphan's court appointed Clark as the child's guardian. Sacagawea's son went on to have a far-ranging career. At age eighteen, he joined a western tour of the young Prince Paul Wilhelm of Württemberg, and afterward went to Europe, where he resided with the prince for six years. The two men returned to America in 1829, and again explored the western country. Jean-Baptiste thereafter was employed as a fur trapper for fifteen years by the American Fur Company. He later served as an army guide during the Mexican War. Joining the gold rush of 1849, Jean-Baptiste set up residence in Placer County, California. Traveling through Montana in May of 1866, he died of pneumonia.

There has been a lively controversy over the correct determination of the date and place of Sacagawea's death. Grace Raymond Hebard, a professor at the University of Wyoming, published the biography *Sacajawea* in 1933, in which she went to great lengths to prove that Sacagawea died April 9, 1884. Hebard traced the alleged wanderings of the "Bird Woman" to the time that she settled down on the Wind River Reservation in Wyoming. Hebard made a substantial case, based on oral testimony of persons who had known the

"Bird Woman"; the hearsay related to known details of the Lewis and Clark expedition. Hebard also relied upon ethnological authorities. At the heart of the controversy is a journal entry of John Luttig, resident fur company clerk at Fort Manuel. On December 20, 1812, he recorded: "this Evening the Wife of Charbonneau, a Snake Squaw died of a putrid fever she was a good and the best Women in the fort, aged abt 25 years she left an infant girl." It is known that Sacagawea had given birth to a daughter, Lizette. The Luttig journal was not published until 1920. Hebard claimed that the death notice referred to Charbonneau's other Shoshone wife, Otter Woman. The issue, however, seems put to rest by the discovery in 1955 of a document in William Clark's journal dated to the years 1825 to 1828. Clark's list of the status of members of his expedition states: "Se car ja we au Dead." Nevertheless, the notion that Sacagawea lived until the 1880's continues to have support.

Impact

Sacagawea had a fourfold impact on the LEWIS AND CLARK Expedition. Though she viewed much of the country the group traversed for the first time, her geographical knowledge was most important in locating the Shoshones in the Rocky Mountains and directing Clark's party through the Bozeman Pass. At crucial instances her services as a translator were essential, and she served as a contact agent. Perhaps, most of all, as an Indian mother with a young baby, she dispelled many of the fears of the Indians encountered on the journey, particularly the fear that the expedition might harm them. She may be credited as a primary factor in ensuring the success of the Lewis and Clark Expedition. Sacagawea also contributed to the uplifting of morale. Throughout the venture she exhibited courage, resourcefulness, coolness, and congeniality. The presence of mother and baby encouraged a certain civilized restraint among the members of the party. Henry Brackenridge, who met Sacagawea in April of 1811, said that she was "a good creature, of a mild and gentle disposition." Clark expressed regrets at the end of the expedition that no special reward could be given to Sacagawea. In many ways she was more valuable to the expedition than her husband, who ultimately received compensation for their efforts.

Sacagawea's place in history was long neglected. Interest in her life, however, gained momentum with the centenary celebrations of the Lewis and Clark Expedition in the early 1900's and especially with the rise of the suffrage movement, which saw in Sacagawea a person of womanly virtues and independence. Eva Emery Dye's novel, *The Conquest: The True Story of Lewis and Clark* (1902), did much during the course of its ten editions to popularize an exaggerated role of Sacagawea on the famous journey of discovery.

Bibliography

Anderson, Irving. "A Charbonneau Family Portrait." *American West* 17 (Spring, 1980): 4-13, 58-64. Written for a popular audience, this article provides a thorough and reliable account of the lives of Sacagawea, her husband Toussaint, and her son Jean-Baptiste.

_____. "Probing the Riddle of the Bird Woman." *Montana: The Magazine of Western History* 23 (October, 1973): 2-17. A scholarly article that persuasively disputes the evidence gathered by Grace Hebard to argue that Sacagawea lived to be nearly one hundred years old.

Brown, Marion Marsh. *Sacagawea: Indian Interpreter to Lewis and Clark*. Chicago: Childrens Press, 1988. A biography written for a juvenile audience.

Chuinard, E. G. "The Actual Role of the Bird Woman." *Montana: The Magazine of Western History* 26 (Summer, 1976): 18-29. Emphasizes the role of Sacagawea as a guide and contact agent and challenges the exaggeration of her actual accomplishments.

Clark, Ella E., and Margot Edmonds. *Sacagawea of the Lewis and Clark Expedition*. Berkeley: University of California Press, 1979. Includes discussion of Sacagawea's life and the efforts made to popularize her legend. Although they provide a relatively accurate account, the authors choose to accept the discredited theory that Sacagawea lived until 1884.

Howard, Harold P. *Sacajawea*. Norman: University of Oklahoma Press, 1971. A balanced biography aimed at a general audience, this work attempts to sort out fact from legend in the life of Sacagawea.

Jackson, Donald, ed. *Letters of the Lewis and Clark Expedition, with*

Related Documents, 1783-1854. 2d ed. 2 vols. Urbana: University of Illinois Press, 1978. Contains a variety of letters, journal entries, and other papers relevant to the activities of the expedition. Sheds some light on the contribution of the Charbonneau family.

Kessler, Donna J. *The Making of Sacagawea: A Euro-American Legend*. Tuscaloosa: University of Alabama Press, 1996. An examination of the development of Sacagawea as a folkloric figure. Includes a bibliography.

Ronda, James P. *Lewis and Clark Among the Indians*. Lincoln: University of Nebraska Press, 1984. This scholarly study examines the contact made between the Lewis and Clark expedition and the Indians. Provides insights into Sacagawea's contributions to the success of the expedition. Includes an appendix that evaluates various books and articles about Sacagawea.

White, Alana. *Sacagawea: Westward with Lewis and Clark*. Springfield, N.J.: Enslow, 1997. Written for a juvenile audience, this volume profiles Sacagawea, with an emphasis on her journey with Lewis and Clark. Includes illustrations, a map, and a bibliography. Part of Enslow's Native American Biographies series.

Harry M. Ward

JEDEDIAH STRONG SMITH

Born: January 6, 1799; Jericho (Bainbridge), New York
Died: May 27, 1831; near Cimmaron River en route to Santa Fe, New Mexico

The most adventurous of the nineteenth century mountain men, Smith charted trails through the Rockies that opened the American West to settlement by the pioneers who followed the fur traders.

Early Life

Jedediah Strong Smith, possibly the greatest of the American mountain men, was born January 6, 1799, in New York State. Like many other American families, the Smiths moved a number of times as the frontier pushed farther west.

Smith's family appears to have been thoroughly middle class and respectable. Smith received a good education and was well read. Family tradition, in fact, credits a book, the 1814 publication by Nicholas Biddle of *History of the Expedition of Captains Lewis and Clark*, with firing the young Smith's imagination and making him determined to see the places Merriwether Lewis and William Clark described in their journals. By 1822 he and his family had made their way to Missouri. There he signed on with the Rocky Mountain Fur Company, recently organized by William Ashley and Andrew Henry. Smith became one of the original Ashley Men, the individual fur trappers and traders that set off into the wilderness under Ashley's command.

Life's Work

Ashley, a Missouri businessman, and Henry, an experienced fur trapper, originated the annual trappers' rendezvous in the intermountain regions of the West. The Rocky Mountain Fur Company would pack supplies and trade goods for the mountain men in to a central location, such as a site on the Green River in Wyoming, and pack the furs out, eliminating the lengthy trek to St. Louis for individual trappers and traders. Ashley's plan met resistance from the Arikara, a Native American tribe whose members had become accustomed to serving as the middlemen between white fur traders and other tribes on the upper Missouri, but Ashley simply relocated his base of operations and effectively cut the Arikara out of the fur business. Other trading companies quickly copied the idea and, before overtrapping ended the fur trade, hundreds of mountain men would gather for the midsummer rendezvous to dispose of the furs taken in the previous year and to stock up on supplies for the coming winter.

Unlike the stereotypical image of the uncouth mountain man, generally portrayed in popular culture as hard-drinking and vulgar in language and behavior, Smith was a devout Christian, neither drank nor smoked, and was consistently serious in his demeanor. Even when personally in danger or in pain, he remained calm, never allowing his men to see his concern or fear. On his second expedition, a grizzly bear attacked and mauled Smith. The bear smashed Smith's ribs and tore at his scalp. The bear left Smith alive, but with his scalp literally dangling by an ear. He coolly instructed one of his companions, Jim Clyman, to reattach the loose skin using needle and thread. Clyman stitched the scalp back in place as best he could but was convinced that repairing the ear was hopeless. Smith told him to try anyway. Clyman did. After a two-week convalescence, Smith resumed command of the party.

Contemporaries of Smith described him both as highly respectable and as an inspiration to those around him. When called upon to say a few words over the grave of John Gardner, a recently deceased fellow fur trader, Smith's eulogy moved observers to comment that Smith left no doubt in anyone's mind that their friend had found salvation. Coupled with his legendary physical courage, it is not surprising Smith quickly established himself both as a leader and as an explorer.

Smith spent his first winter in the mountains along the Musselshell River in present-day Montana. The following summer, 1823, Ashley directed Smith to take a group of men and find a Native

American tribe known as the Crow with the object of establishing trade relations with them. Smith's party succeeded in making contact with the Crow people and spent the winter with them. Members of the Crow tribe described to Smith the Green River area in what is now the state of Wyoming. The Green was reputed to be an area rich in furs and thus far unexploited by other traders, and Smith and his men resolved to explore the region. In the spring of 1824 Smith and his party rediscovered South Pass, a passageway through the Rocky Mountains, and successfully traversed it with both wagons and livestock, proving that such travel was possible. Previous parties had relied completely on pack animals to carry supplies and trade goods. South Pass had been utilized before by white men, in 1814, but the route had been forgotten. Other trails, such as the one over Lemhi Pass used by the LEWIS AND CLARK Expedition, were impassable with wagons and were just barely passable with horses and mules. The trail that Smith blazed through Wyoming in 1824 later became an integral part of the Oregon Trail, the path that thousands of pioneers would take to reach the Pacific Northwest.

Smith and the men with him spent the following year trapping and trading in Wyoming and Idaho. Following the first trappers' rendezvous, he returned to St. Louis with William Ashley and the company's furs. Andrew Henry had decided to retire from the fur trade, so Ashley asked Smith to replace Henry as his partner.

Smith returned to the mountains ahead of Ashley and his main party the following spring to arrange for the trappers' rendezvous. At the rendezvous, Ashley negotiated the sale of his share of the company to Smith and two new partners, David Jackson and William Sublette. Ashley—in exchange for a promissory note signed July 18, 1826, which committed Jackson, Smith, and Sublette to pay "not less than seven thousand dollars nor more than fifteen thousand dollars" for merchadise-agreed to arrange for the shipment of trade goods to the location of the following year's rendezvous. After the rendezvous ended, Ashley returned to St. Louis while Smith, Jackson, and Sublette divided their party into smaller groups for the fall hunt.

Smith, accompanied by seventeen men, decided to explore the

region south of the Great Salt Lake and to assess its potential for the fur trade. Smith and his men traveled the length of Utah, following first a tributary of the Colorado River and then Colorado itself, pushed on into what is present-day northern Arizona, and then crossed the Mojave Desert into California, eventually reaching the Spanish mission at San Gabriel near present-day San Diego. The 1776 Spanish Dominguez-Escalante Expedition had attempted this route across the desert but had failed to complete it. The bulk of Smith's party remained on the Stanislaus River in California in the spring of 1827 while Smith and two men attempted to find a route back to northern Utah through the Sierra Nevada. They succeeded and, striking northeast across Nevada, became the first white men to cross the Great Salt Lake Desert as they returned to Utah for the trappers' rendezvous at Sweet (now Bear) Lake.

Smith, as was common practice among many American explorers, fur traders, and mountain men, kept extensive journals. His harrowing description of the journey across the Salt Lake Desert—replete with phrases such as "I durst not tell my men of the desolate prospect ahead" and "We dug holes in the sand and laid down in them for the purpose of cooling our heated bodies"—makes it clear that even Smith had his doubts regarding their survival. Nonetheless, having survived the trek across not merely one but several arid deserts, Smith continued his explorations. He arrived at the rendezvous July 3, having traveled through most of the American Southwest during the previous year, only two days later than he had promised Ashley in 1826 that he would be there. Scarcely two weeks after completing his harrowing journey from California, he was again heading south and west, motivated, as he said in his journal, "by the love of novelty."

Having left a significant number of his men in California in the spring of 1827, Smith retraced his route to the Pacific Ocean in the fall of that year. Highly suspicious of Smith's motives, the Spanish governor threatened the Americans with jail. Officials softened their stance and did allow Smith and his men to spend the winter of 1827-1828 in the San Francisco Bay area, but they made it clear that they did not want the Americans to linger any longer than necessary. In the spring they proceeded north to present-day Oregon, and Smith became the first white man to travel from California to Oregon by an overland route. The Kelawatset Indians of the region proved hostile, however, and killed the majority of Smith's men in an attack. Smith and three other survivors managed to reach Fort Vancouver, where they were aided by British trappers. After spending the winter of 1828-1829 at Fort Vancouver, Smith returned to the Flathead region for the 1829 trappers' rendezvous. Briefly reunited with his partners, Jackson and Sublette, Smith then led a large force of men into the Blackfoot country of Montana and Wyoming for the fall hunt.

The Indians of the northwest were becoming increasingly unfriendly, worsening the risks to both trappers and traders, so in 1830 Sublette, Smith, and Jackson decided to sell their trapping interests to the Rocky Mountain Fur Company. They returned to St. Louis and became involved in the growing trade with Santa Fe. Smith himself planned to give up the wandering life and settle down in St. Louis. His mother had died recently, and he felt a strong sense of obligation to his family. By 1830 he had spent eight years in almost constant travel and exploration. Perhaps the novelty of new places was finally losing some of its allure. He purchased both a farmhouse and a town house, hired servants, and talked about preparing his complete journals and maps for publication. Still, he allowed himself to be persuaded to make one last trip. In the spring of 1831 he agreed to lead a trading expedition to Santa Fe to help the buyers of his fur company procure supplies. A band of Comanches apparently surprised and killed Smith while he was scouting ahead of the main party in search of drinking water near the Cimmaron River along the Santa Fe Trail. The planned editing and publication of his complete journals never took place, and most of his papers were lost following his death.

Impact

Although numerous American explorers charted sections of the continent, few covered as much territory or saw as wide a variety of terrain as Smith, nor did their travels have as significant an impact on later settlement. Smith's journal, written on his trek from the

Green River in Wyoming to Arizona and then on to the Pacific coast, contains the first descriptions by Americans of both the wonders of the Grand Canyon and the magnificent redwood groves of California. His trek across the South Pass of the Rockies, a five-hundred-mile journey with pack wagons and livestock, opened a trail that would be utilized by thousands of pioneers en route to Oregon. Similarly, his trek across the Great Salt Lake Desert and Nevada blazed a more direct route to California. It later served as the route first for the Pony Express and then for U.S. Highway 50.

Bibliography

Allen, John Logan. *Jedediah Smith and the Mountain Men of the American West.* Introduction by Michael Collins. New York: Chelsea House, 1991. This biography written for a juvenile audience chronicles the exploits of the early nineteenth century mountain men who opened trails through the American West. Bibliographical references and index.

Brooks, George R., ed. *Southwest Expedition of Jedediah S. Smith: His Personal Account of the Journey to California, 1826-1827.* Lincoln: University of Nebraska Press, 1990. Smith's journey across the Mojave and the Great Salt Lake Desert, in his own words.

Dale, Harrison Clifford. *The Explorations of William H. Ashley and Jedediah Smith, 1822-1829.* Lincoln: University of Nebraska Press, 1991. Originally published as *The Ashley-Smith Explorations and the Discovery of a Central Route to the Pacific, 1822-1829,* in 1941. Includes original journals edited by Dale. Excellent history that summarizes the travels of both Ashley and Smith and sets them in a historical context.

Davis, Lee. "Tracking Jedediah Smith Through Hupa Territory." *The American Indian Quarterly* 13 (Fall, 1989): 369. Provides vivid details about one segment of Smith's travels. Davis, by looking at one aspect of Smith's explorations in detail, helps to flesh out the more general accounts of his travels.

Morgan, Dale L. *Jedediah Smith and the Opening of the West.* Lincoln: University of Nebraska Press, 1994. A good basic biography of Smith, containing a portrait.

Neihardt, John Gneisenau. *The Splendid Wayfaring: Jedediah Smith and the Ashley-Henry Men, 1822-1831*. Lincoln: University of Nebraska Press, 1990. Fascinating examination of Smith and his fellow fur traders and trappers and their mythic status in American history.

Smith, Jedediah Strong. *The Southwest Expedition of Jedediah S. Smith: His Personal Account of the Journey to California, 1826-1827*. Edited by George R. Brooks. 1977. Reprint. Lincoln: University of Nebraska Press, 1989. These accounts, by the explorer himself, are supplemented by a bibliography and an index.

Sullivan, Maurice S. *The Travels of Jedediah Smith*. Omaha: University of Nebraska Press, 1992. Originally published in 1934 using materials from Smith's surviving journals, this book has long been considered the definitive reference on Smith's life and travels.

Nancy Farm Mannikko

JOHN SMITH

Born: January 9, 1580 (baptized); Willoughby, England
Died: June 21, 1631; London, England

Smith's strong leadership in early Virginia and his promotional literature on North America helped ensure the success of England's efforts at colonization.

Early Life

John Smith was born in January, 1580, in the small Lincolnshire village of Willoughby. His father, George Smith, a yeoman farmer, and his mother, née Alice Rickard, were from families long linked to the land in northern England. Smith was the eldest of the couple's five sons and one daughter. He attended local schools until age fifteen, primarily studying grammar and mathematics, and then was apprenticed to a merchant in the coastal town of King's Lynn. Though he inherited some land after his father's death in 1596, Smith joined with other English Protestants in their conflict with the Spanish in the Netherlands. After three years of fighting, Smith rambled about France and Scotland, at times as a companion to the sons of his noble

Archive Photos

patron, Lord Willoughby, and at times as a solitary wanderer.

These episodes prompted the restless young man to seek new adventures. Throughout his life, he was eager to experience the demanding and the dangerous. Motivated in part by the thrill of the challenge, Smith, as a commoner, also sought to prove his worth in an age dominated by gentlemen. Accordingly, Smith decided to take up the seemingly endless Christian crusade against Islam. In 1600, he started across the continent to fight the infidel Turks in Hungary. Over the next four years, Smith engaged in privateering on the Mediterranean Sea as well as in numerous battles in Hungary, suffering serious wounds, capture, and enslavement; after he escaped, Smith trekked through Russia, Poland, Germany, France, and Spain before returning to England. A 1616 portrait of Smith reveals a short and stocky adventurer in military garb. His face features a high forehead, a long, slender nose, and a full beard. The artist captured the confident air which characterized Smith throughout his career.

Life's Work
In 1606, when Smith learned that the Virginia Company of London intended to settle a colony in North America, he enlisted for what became the most important venture of his life. The approximately 105 men dispatched by the joint stock company in three small ships established a base on the James River in Virginia in the spring of 1607. The company appointed a resident council to supervise affairs in the colony, and Smith was one of the seven selected. During his two and a half years at Jamestown, Smith was instrumental in furthering the survival of the colony. He led numerous exploratory and mapping ventures along the coast and into the interior. He established a vital trade with the leader of the Tidewater natives, Powhatan. Smith also struck up an important, paternal friendship with the chief's daughter, Pocahontas, a youngster who served as a liaison between the natives and the English. Her role in promoting the trade of food with the colonists led Smith later to claim that she saved them "from death, famine, and utter confusion."

When Smith was elected president of the council in September, 1608, he faced a host of problems. Most of the initial settlers, almost half of whom were gentlemen, were poorly suited to the harsh task of creating a settlement in a hostile environment. Further, several settlers refused to work because they had assumed incorrectly that they, like the Spanish in Mexico and Peru, would be able to compel the natives to do the hard labor. Even those colonists who were willing to work could not contribute because they were often ill (or had died) from malaria or dysentery, a consequence of building Jamestown on swampy ground. Indian attacks, although sporadic, also claimed some lives. The other leaders could provide little guidance since the council suffered from unending dissension. Yet, to Smith, the most significant dilemma was gold fever. Hoping to replicate the Spanish success in discovering New York gold, for most, "there was no talk, no hope, no work but dig gold, wash gold, refine gold, [and] load gold," Smith explained.

Smith quickly imposed the discipline necessary to save the colony. He forced the surviving men to build new housing at Jamestown, enlarge the fort, and plant more than thirty acres of grain. In short, he made them work; if they did not, they did not eat. Smith also secured, through cajolery, intimidation, and brutality, a steadier supply of food from the natives and relatively peaceful relations with them. His significance to the fledgling colony was most evident when a serious gunpowder burn forced Smith to return to England in October, 1609. Shortly after his departure, Virginia once again fell into chaos, and almost ninety percent of the settlers died in the winter.

The documentary record permits only occasional glances into Smith's life after he left the troubled colony. He returned to North America in 1614, to explore and map the New England coast. The experience convinced Smith that colonists could prosper there by exploiting the region's fish and fur. He secured financial backing for two more voyages to New England, but French pirates and storms prevented their completion. He was briefly reunited with Pocahontas when she came to England in 1616. She had married planter John Rolfe, and he had changed her name to Rebecca. It was a bittersweet reunion because she died in early 1617. Two years later, Smith offered his expertise and service to a group of religious dissenters, the Pilgrims, when he discovered that they intended to settle in North America, but they turned down his offer. Smith was also snubbed by the Virginia Company of London. They rejected both his plea for compensation and his offer to lead an armed force against the natives to avenge their assault on the colony in 1622.

Impact

In the last half of his life, Smith became less a man of action and more a writer and advocate of further colonization. In all, he published ten works. These autobiographical sketches, histories, reprints of other accounts, guides for seamen, and promotional tracts comprise the most important source of information on Smith's life. For generations, historians dismissed them as unreliable because Smith's portrayal of his career was so fantastic. He did include episodes which stretch the limits of credibility. For example, he tended to attribute his escapes from difficult circumstances to the intervention of women. When the Turks captured him, Smith explained, they sold him as a household slave to a young woman named Charatza Tragabigzanda. She quickly became attracted to the Englishman and worked to ease his enslavement. After he escaped and made his way across Russia, Smith claimed that a noblewoman there aided him. His most famous rescue, however, occurred in Virginia. After capture by Powhatan's warriors, the eleven-year-old Pocahontas prevented his execution. There were others. As he escaped from the French pirates who had captured his ship headed for New England, Smith wrote that he was helped by a Madame Chanoyes. Finally, when he had difficulty finding a backer for his longest book, *The Generall Historie of Virginia, New-England, and the Summer Isles* (1624), Frances Howard, the Duchess of Richmond and Lennox, provided the funds. Smith also included incredible claims of victories over great warriors in single combat. The most prominent example came while he fought in Hungary. He met on successive days three Turkish opponents and not only killed them but also removed their heads for trophies.

Even Smith's most sympathetic biographers have acknowledged his enormous ego and his determination to emphasize his role in the events he chronicled. (In his publications, he referred to himself as captain, or governor of Virginia, or admiral of New England, or all three.) Nevertheless, Smith's writings remain important. Modern research has largely confirmed the accuracy of his histories (although the episodes involving female saviors and single combat have not been substantiated). Indeed, his *A True Relation of Such Occurrences and Accidents of Noate as Hath Hapned in Virginia Since the First Planting of That Collony* (1608) and *A Map of Virginia, with a Description of the Country* (1612) are the most complete accounts of early Virginia and the best sources on the natives of the Tidewater region. Moreover, Smith's work had significance in his own era. *A Description of New England: Or, Observations and Discoveries of Captain John Smith* (1616), *New Englands Trials* (1620), and *The Generall Historie of Virginia, New-England, and the Summer Isles* served as promotional literature for expanding the English empire. He wrote in glowing terms of the prospects for North America. Yet, to build flourishing colonies, England would have to learn from his experiences in Virginia and New England. Success would come, Smith argued, only from a combination of firm leadership, industrious settlers, and aid from the English crown.

Smith closely followed events in the New World until his death in London in 1631. He lived long enough to learn of Virginia's stability under royal control (King James I revoked the Virginia Company charter in 1624) and of the large Puritan migration to Massachusetts, developments which vindicated his enthusiastic support of empire. The unashamedly ambitious and arrogant Smith had been a crucial figure, both as a leader and as a promoter, in England's efforts to colonize North America.

Bibliography

Adams, Henry. "Captaine John Smith." In *Historical Essays*. New York: Charles Scribner's Sons, 1891. Reprint. Hildesheim, Germany: Georg Olms Verlag, 1973. In this essay, Adams, the great nineteenth century historian and autobiographer, offers some of the harshest criticism of Smith. He not only questions Smith's reliability as a writer but also concludes that Smith's career ended in failure.

Barbour, Philip L. *The Three Worlds of Captain John Smith*. Boston: Houghton Mifflin Co., 1964. This is the most complete biography by the leading authority on Smith. Barbour presents the greatest details on Smith's years in Virginia and the most thorough discussion of sources.

Emerson, Everett H. *Captain John Smith*. New York: Twayne Publishers, 1971. Rather than a biography, this is a study which focuses on Smith as a writer. While acknowledging Smith's embellishments, Emerson praises his prose, particularly his ability to draw vivid word pictures.

Hawke, David Freeman, ed. *Captain John Smith's History of Virginia: A Selection*. Indianapolis: Bobbs-Merrill Co., 1970. Hawke offers an edited reprint of the sections on Virginia in Smith's *The Generall Historie of Virginia, New-England, and the Summer Isles*. London: I. D. and I. H. for Michael Sparks, 1624. While Hawke has a good, brief introduction, he includes scarcely any footnotes to help the reader.

Morgan, Edmund S. *American Slavery, American Freedom: The Ordeal of Colonial Virginia*. New York: W. W. Norton and Co., 1975. Morgan describes the origins of slavery in Virginia and its impact on the colony's economic, social, and political development. The early chapters include an excellent description of the difficulties faced by Smith and the other settlers at Jamestown.

Smith, Bradford. *Captain John Smith: His Life and Legend*. Philadelphia: J. B. Lippincott Co., 1953. In this well-researched volume, Smith receives his most sympathetic treatment. Particularly valuable on John Smith's early years. The author draws upon the research of Laura Polanyi Striker, which largely substantiates Smith's account of his experiences in Hungary.

Vaughan, Alden T. *American Genesis: Captain John Smith and the Founding of Virginia*. Boston: Little, Brown and Co., 1975. Vaughan does more than present a short, balanced biography of Smith. He also details the history of Virginia from Smith's departure in 1609 until his death in 1631.

Larry Gragg

HERNANDO DE SOTO

Born: c. 1496; Jérez de los Caballeros?, Spain
Died: May 21, 1542; near modern Ferriday, Louisiana

After playing a prominent role in the conquest of Nicaragua and Peru, de Soto led the 1539-1542 expedition which explored what became the southeastern United States and discovered the Mississippi River.

Early Life

Hernando de Soto was born around 1496, probably in Jérez de los Caballeros in southwestern Spain. He was the second son of Francisco Méndez de Soto and Leonor Arias Tinoco (the proper family name is Soto but the English-speaking world calls him de Soto). The family was lower nobility, and Hernando received some education, although he was always by temperament a soldier and adventurer.

De Soto could expect to inherit little from his father's small estate and thus sailed to Central America in 1513-1514 with Pedro Arias Dávila (called Pedrarias), the new governor of Panama. Pedrarias allowed his followers to ravage Central America as long as they respected his authority. In these lawless conditions, de Soto flourished. Above average in height, bearded, and darkly handsome, he was vigorous, brave, and aggressive, always in the vanguard. His spoils in land and Indian workers made him wealthy.

By 1517, de Soto was a captain. He soon formed a partnership with Hernán PONCE DE LEÓN and Francisco Campañón to share equally in the booty which fortune might bring them. They helped conquer Nicaragua in 1524. Campañón died in 1527, but de Soto and Ponce de León stayed on in Nicaragua. De Soto served a year as magistrate of León, although temperamentally unsuited to administration. He also became ambitious for his own governorship, but Pedrarias blocked those aspirations in Nicaragua.

Life's Work

In 1530, de Soto agreed to join forces with Francisco PIZARRO in the conquest of Peru. When he met Pizarro at Puná Island in late 1531, de Soto had two ships, one hundred men, and some horses. He expected to be Pizarro's second in command and to receive an independent governorship. Pizarro gave de Soto charge of the vanguard. On a scouting foray inland to Cajas, de Soto seized several hundred Indian women and turned them over to his female-starved men. The Spaniards also learned at Cajas that the Inca Empire had been torn apart by a great civil war and that Atahualpa, leader of the victorious faction, was encamped with his army at Cajamarca not too far away.

When the Spaniards arrived at Cajamarca, Pizarro sent de Soto with a small detachment to greet Atahualpa. A great horseman, de Soto tried unsuccessfully to frighten the emperor, who had never seen a horse before, by riding right up to him: Atahualpa accepted Pizarro's invitation to visit the Spaniards in Cajamarca the next day. Once he was inside the city walls, they took him captive. While a hostage, Atahualpa became close to de Soto and gave him valuable gifts. De Soto was on a scouting expedition when Pizarro executed Atahualpa. Upon his return, de Soto criticized the execution, arguing that Atahualpa should have been sent to Spain as a prisoner rather than killed. Always punctilious about keeping bargains, de Soto was also upset that Pizarro had killed the emperor after Atahualpa had filled rooms with gold and silver, as his agreed-upon ransom.

Archive Photos

On the march from Cajamarca to the Inca capital, Cuzco, during the second half of 1533, de Soto again led the vanguard. As they neared Cuzco, he disregarded orders and rushed ahead with his small force to claim credit for occupying the city. The Indians ambushed his party at Vilcaconga, however, and only the timely arrival of reinforcements saved him. PIZARRO appointed de Soto lieutenant governor of Cuzco in 1534 but replaced him by that year's end. Convinced that Pizarro would never give him an independent command in Peru, de Soto headed for Spain. He left behind Leonor Curuilloi, an Inca princess and his mistress, and their daughter Leonor. PONCE DE LEÓN came from Nicaragua to manage de Soto's property.

In Spain by 1536, de Soto had 100,000 pesos with him, a reputation as a great conqueror and explorer, and a hunger to lead a new expedition. He petitioned the king. On April 20, 1537, Charles V made him governor and captain general of Cuba and Florida and gave him authority to explore and conquer the New Land at his own expense. Álvar Núñez CABEZA de Vaca, a survivor of the disastrous Panfilo de Narváez expedition to Florida, had come to Spain with tales of great riches in the New Land. De Soto tried but failed to persuade him to join the expedition. Hundreds of young adventurers rushed to enlist, however, assuming that Florida would make them rich and famous. His wife, Doña Isabel de Bobadilla (Pedrarias' daughter, whom he had married in 1536), and about seven hundred carefully chosen soldiers of fortune sailed with him to Cuba in 1538.

De Soto spent a year in Cuba to train his men and gather provisions. The expedition departed for Florida in May, 1539, with de Soto's wife remaining behind to govern Cuba. It landed at Espiritu Santo (Tampa) Bay on May 30. At the outset, de Soto discovered Juan Ortiz, a survivor of the Narváez expedition who had lived with the Indians for twelve years. Ortiz became de Soto's interpreter.

De Soto had made his fortune in Central America and Peru by plundering the Indians. He intended to do the same in Florida. When the expedition came to an Indian settlement, de Soto took the chief hostage so that the Indians would serve the Spaniards while they ravaged the village. When the expedition was ready to move on,

DE SOTO'S EXPEDITION, 1539-1543

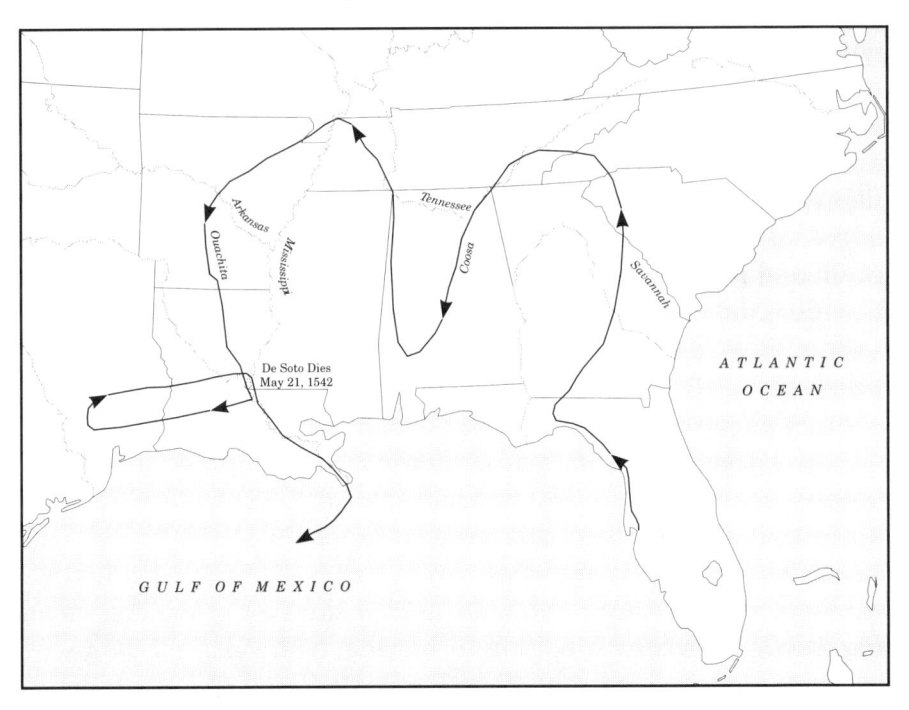

de Soto forced the hostage chieftain to provide porters (he took neck irons and chains to Florida with this aim in mind). De Soto released the porters and chief at the next village, where new hostages were seized. Those who resisted were mutilated, burned alive, or thrown to the dogs.

The Spaniards wandered through what later became Florida, Georgia, South Carolina, North Carolina, Tennessee, Alabama, Mississippi, Arkansas, Louisiana, and Texas. De Soto knew nothing of the land, nor did he have a destination in mind. He simply assumed that the Indians of the New Land would have riches to plunder. The expedition proceeded north from Tampa Bay and soon exhausted its supplies. From then on, the explorers lived off the land and Indian agriculture, plus a herd of swine which de Soto had brought in the fleet. Turning west, they reached the rich agricultural lands of the Apalachees in the Florida panhandle and stayed from October to March, 1540.

Then they marched northeast toward Cofitachequi, on the Savannah River near what is now Augusta, Georgia. Their porters had died during the winter, and the Apalachees had run away before they could be enslaved. Thus, the Spaniards had to carry their own food and equipment. At Cofitachequi, they found large quantities of freshwater pearls, and de Soto himself obtained a chestful after plundering the village and burial grounds. Taking the queen hostage and thus provided with porters, de Soto headed northwest through Cherokee land in the Carolina piedmont and then turned west toward the Tennessee River. The Spanish learned about tribes rich in gold somewhere farther on but could never find them. De Soto cared nothing about mining; only plunder interested him.

In July, 1540, de Soto marched south, having learned of a rich people called Coosa (Creek territory). They were moving through Alabama, toward Mobile Bay. Disappointed by Coosa, they continued southwest and came to the Mabilas (Choctaws). At Mabila, the Indians revolted, killed twenty-two Spaniards, and burned much of the equipment, including de Soto's pearls. By then, more than one hundred of the men who had started with de Soto at Tampa Bay were dead. Although not too far north of Mobile Bay, de Soto refused to push on to the coast, fearful that his men would desert. Instead, he turned north again and occupied a Chickasaw village for winter quarters. After enduring the Spaniards for several months, the Chickasaws revolted on March 4, 1541. They set fire to the village and killed a dozen Spaniards, fifty horses, and many of the pigs. The Spanish escaped annihilation only because a sudden storm prevented the Indians from immediately renewing their attack.

Pushing westward, de Soto discovered the Rio Grande (Mississippi) on May 8, 1541, built barges, and crossed the mighty river on June 18 a little above its junction with the Arkansas. The Spaniards wandered for two months through central Arkansas. De Soto sent a scouting party to the Ozarks after rumors of gold there. They spent the severe winter of 1541-1542 at Utiangue in south-central Arkansas.

Battered and discouraged, the expedition left Utiangue on March 6, 1542, and followed the Ouachita River south through

Louisiana to the Mississippi. De Soto fell ill with a fever. Realizing that his end was near, he named Luis de Moscoso, another veteran of Peru, to succeed him. The explorer died on May 21, 1542, to the relief of those who wanted to abandon the quest. His men sank his body in the Mississippi to hide it from the Indians.

The men decided to march overland to Mexico and traveled several hundred miles into Texas. After four months and no sign of Mexico, they turned back. They spent their last winter near the Mississippi at Aminoya (northwest of Natchez). In the spring of 1543, they built barges, and, floating down the river to the Gulf of Mexico in July, they then sailed west. Clad in rags and animal skins when they arrived in Mexico in September, 1543, about one third survived of those who had started at Tampa Bay.

Impact

De Soto evokes conflicting impressions. On the one hand, he was certainly one of the bravest Spaniards of his time. He was gallant and courageous, the epitome of the explorer and conquistador. On any expedition, he was always in the vanguard; in any battle, he was in the front ranks. He amassed a huge fortune in Nicaragua and Peru, and he failed to do so in North America only because there were no rich Indian cities to plunder. Yet adventure, danger, and the unknown seemed to attract him more than riches. His expedition was extremely well organized. He recruited not only soldiers but also artisans, who could build boats and bridges. He raised a herd of pigs on the march so that the expedition would have meat later. Despite rugged terrain and many attacks from fierce Indians, he held the expedition together.

On the other hand, de Soto was a plunderer, not a builder. His expedition made no effort to settle or colonize; nor did it even attempt to exploit the natural resources of the region. De Soto had learned too well the lessons under Pedrarias, and the experiences in Peru reinforced them. Perhaps he took less delight in butchery than some Spaniards, but he was ready to torture and kill in his quest for the riches of Florida. Indians' lives were worth little to him. Although he considered Christianization of the Indians one of his expedition's responsibilities, he did little to achieve it. Moreover, he was too stubborn and proud to end the foray despite its obvious failure to find the booty for which the Spaniards hunted.

The Spanish atrocities should not, however, obscure the achievements of the expedition. It was the first major European penetration of the North American interior and left valuable information about the Indians and geography of the region. De Soto and his men left a legacy of courage, ambition, and perseverance rarely equaled as they opened the region to European expansion. The great river they discovered eventually proved to be a natural treasure more valuable than the booty de Soto sought.

Bibliography

Blanco Castilla, Francisco. *Hernando de Soto, el centauro de las Indias*. Madrid: Editorial Carrera del Castillo, 1959. Based on the major documentary sources but too novelistic and heroic in tone to be totally satisfactory.

Bourne, Edward Gaylord, ed. *Narratives of the Career of Hernando de Soto*. 2 vols. New York: Allerton Book Co., 1904. Contains English translations of the three most important chronicles of de Soto's expedition by Luis Hernández de Biedma, factor of the expedition, Rodrigo Ranjel, de Soto's secretary, and an unidentified Portuguese gentleman from Elva.

Duncan, David Ewing. *Hernando de Soto: A Savage Quest in the Americas*. New York: Crown Publishers, 1995. This illustrated volume contains maps, bibliographical references, and an index.

Galloway, Patricia, ed. *The Hernando de Soto Expedition: History, Historiography, and "Discovery" in the Southeast*. Lincoln: University of Nebraska Press, 1997. Includes a bibliography and an index.

Garcilaso de la Vega, The Inca. *The Florida of the Inca*. Translated and edited by John Grier Varner and Jeannette Johnson Varner. Austin: University of Texas Press, 1951. The most famous of the four early chronicles of de Soto's expedition. Unlike the others, it was not written by an eyewitness and is thus more problematical.

Hemming, John. *The Conquest of the Incas*. New York: Harcourt Brace Jovanovich, 1970. An excellent, well-written account of the conquest of Peru, with information about de Soto's role in it.

Hudson, Charles M. *Knights of Spain, Warriors of the Sun: Hernando de Soto and the South's Ancient Chiefdoms*. Athens: University of Georgia Press, 1997. Hudson's illustrated work offers maps, an extensive bibliography, and an index.

Maynard, Theodore. *De Soto and the Conquistadores*. New York: Longmans, Green and Co., 1930. The best biography of de Soto, although marred by inaccuracies and omissions.

Milanich, Jerald T., ed. *The Hernando de Soto Expedition*. New York: Garland Publishers, 1991. Includes narratives of the career of de Soto as told by a Knight of Elvas, the diary of Rodrigo Ranjel, and a Canete fragment.

Milanich, Jerald T., and Charles Hudson. *Hernando de Soto and the Indians of Florida*. Gainesville: University Press of Florida, 1993. Includes illustrations, maps, bibliographical references, and an index.

Ober, Frederick A. *Ferdinand de Soto and the Invasion of Florida*. New York: Harper and Brothers, Publishers, 1906. Based extensively on Garcilaso's aforementioned narrative and aimed primarily at younger readers.

Oviedo y Valdés, Gonzalo Fernández de. *Historia general y natural de las Indias*. 5 vols. Madrid: Ediciones Atlas, 1959. A sixteenth century history of the New World. Book 12 narrates de Soto's expedition to Florida. Drawn from Ranjel's account.

Sauer, Carl Ortwin. *Sixteenth Century North America: The Land and the People as Seen by the Europeans*. Berkeley: University of California Press, 1971. Contains a chapter on de Soto's expedition by the leading historical geographer of sixteenth century North America. Critical of de Soto's motives and behavior.

Sola y Taboada, Antonio del, and José de Rújula y de Ochotorena. *El Adelantado Hernando de Soto: Breves noticias, nuevos documentos para su biografía*. Badajoz: Ediciones Arqueros, 1929. Particularly important for its documentary appendices, which include the agreement between de Soto and Ponce de León, de Soto's capitulation for the conquest of Florida, the information about his background which he submitted to enter the Order of St. James, his will, and an inventory of his property.

Soto, Hernando de. *The de Soto Chronicles: The Expedition of Her-

nando de Soto to North America in 1539-1543. 2 vols. Edited by Lawrence A. Clayton, Vernon James Knight, Jr., and Edward C. Moore. Tuscaloosa: University of Alabama Press, 1993. Illustrated with maps, bibliography, and an index.

United States de Soto Expedition Commission. *Final Report of the United States de Soto Expedition Commission*. Washington, D.C.: Government Printing Office, 1939. A scholarly, definitive study of the route followed by the de Soto expedition.

Kendall W. Brown

JOHN HANNING SPEKE

Born: May 4, 1827; Orleigh Court, Devon, England
Died: September 15, 1864; Neston Park, near Bath, England

Speke traveled extensively in East and Central Africa and during the course of his explorations discovered Lake Victoria, the source of the Nile River.

Early Life

John Hanning Speke was born May 4, 1827, at Orleigh Court in southwestern England. He was the son of William and Georgiana Hanning Elizabeth Speke. His father was a retired army officer, and his mother came from a family of wealthy merchants. Relatively little is known of Speke's early years. He showed an interest in zoology from a tender age but was a restless boy who cared little for school. Whenever possible, he was out in the fields and woods, and it is from these youthful days that a lifelong interest in natural history and sport dates.

In 1844, the same year his father came into possession of the family estate at Jordans (many sources wrongly suggest that this was Speke's birthplace), Speke received assignment as a second lieutenant in the Forty-sixth Bengal Native Infantry Regiment, in the Indian army. This assignment followed completion of his studies at Blackheath New Preparatory School in London.

April 24, 1844, may be said to be the day when Speke was vested with the responsibilities of manhood. On this day the seventeen-year-old lad was examined and passed as an entering officer cadet for Indian army service. Speke's preferment was in all likelihood largely a result of his mother's influence with the Duke of Wellington. He had already passed the required medical examinations prior to his final application, so he was able to begin active service almost immediately. On May 3, 1844, he boarded ship and four months later was in India. The Asian subcontinent and its life apparently suited him well. He demonstrated some facility with languages, and by the end of his second year of service, he had passed an examination in Hindustani. He also had ample opportunity to indulge his love of hunting and made many forays into the plains and the Himalaya Mountains in search of sport.

Speke saw active service in the Second Sikh War when he served as a subaltern officer in the "fighting brigade" of General Colin Campbell's division, and on October 8, 1850, he was promoted to the rank of lieutenant. It was also at this point that he first began to think of African exploration, although at the time collecting specimens of the continent's animal life weighed more heavily in his mind than geographical discovery. For the time, though, any such journey was only a dream. He still had five years' service remaining before he would become eligible for an extended furlough.

Meanwhile, he scrupulously saved his money and underwent conscious training in preparation for African exploration. Each year he made excursions into the mountains of Tibet, he developed his skills as a surveyor and cartographer, and he became an excellent hunter and marksman. Speke actually cared little for the dull routine of army life, as he later confessed, and he clearly was a man in search of adventure. This obsession would drive him all of his life, as he constantly sought to conquer that which was mysterious or unknown. This trait, coupled with his interest in nature, was vital in his choosing the arduous life of an explorer.

Thanks to the favor of his superiors, he was able to obtain regular extended leaves, and late in 1854, he had the opportunity for which he had so long waited. He was to join a fellow Indian army officer, Richard Francis BURTON, in exploring Somaliland. He and Burton had met on shooting outings in India, and their common interest in Africa had drawn them together. From this juncture onward, for the remaining ten years of his life, Speke would be almost completely preoccupied with African discovery.

Life's Work
The Somaliland undertaking proved an abortive one. While camped on the coast, the party was attacked, with Speke being badly

wounded and briefly held captive. The attack also planted the seeds of discord between BURTON and Speke, for in the confusion Burton believed that his companion had not responded to the native attack as readily or bravely as he should. Yet realizing that Speke had "suffered in person and purse," Burton invited him to join another expedition. This journey began in December, 1856, with its primary objective being the discovery of the mysterious sources of the Nile River.

The pair, rather than taking the standard approach of moving upriver along the Nile, traveled overland from the East African coast opposite Zanzibar. Together they discovered Lake Tanganyika, which would ultimately prove to be the source of the Congo River, although Burton thought it might be their objective. On the return journey, while Burton remained in camp at Kazeh (modern Tabora, Tanzania) investigating some of the social and sexual customs which so fascinated him, Speke made a flying march to the north.

His objective was a lake which local reports said stretched to the ends of the earth, and on July 30, 1858, he first sighted the vast body of water which he instinctively knew was the Nile's source. He named it Lake Victoria, after his sovereign, but because of time considerations and the fact that he was nearly blind from ophthalmia, Speke had no opportunity for a proper reconnaissance of the lake. Nevertheless, he returned to Kazeh proclaiming that he had discovered the Nile's source, a conclusion which BURTON ridiculed. Henceforth there would be bitter discord between the two, although they had little choice but to retain at least the vestiges of friendship during the perilous return journey to Zanzibar.

Once back in England, however, Speke immediately claimed credit as the discoverer of the Nile's source, and his theories attracted widespread attention. He had preceded Burton back to England, and when the leader of the East African Expedition returned a few weeks later, he found, to his regret, that Speke was the "lion of the day." The Royal Geographical Society, with the encouragement of Sir Roderick I. Murchison, had decided to finance a new expedition for the purpose of obtaining proof of Speke's claims.

Accompanied by James Augustus Grant, another acquaintance from his Indian army days, Speke returned to East Africa in 1860.

Despite facing an incredible variety of obstacles, ranging from incipient warfare among tribes living along the route to recurrent demands for *hongo* (passage fees), the expedition made its way to the important lake kingdoms of Buganda and Karagwe. While Grant stayed behind at the latter location (ostensibly because of a leg injury but probably owing to Speke's egotistical reluctance to share any fame which might come from his discoveries), Speke marched eastward to the Nile. On July 21, 1862, he reached the river and followed it upstream until he reached the falls at the north end of Lake Victoria. This massive outflow of water he named the Ripon Falls, in honor of the president of the Royal Geographical Society, Lord Ripon.

At this juncture, his task seemed simple enough. All that remained was to collect Grant and proceed downriver until the expedition reached a known point on the Nile. This proved impossible though, thanks to objections raised by tribes living along the river. The necessity of taking a detour away from the Nile as he moved downriver left Speke's claims open to dispute, and it would not be until 1890, long after his death, that his contentions would be proved completely correct. Nevertheless, after overcoming a series of vexing impediments placed in their way by African chieftains, Speke and Grant managed to resume their travels downriver. Falling in with a company of Arab slave and ivory traders, they reached Gondokoro, the last outpost of European civilization, on February 15, 1863.

There they met Samuel White Baker and his mistress (later to become his wife) Florence von Sass. After a few days with this extraordinary couple, they proceeded onward in their journey back to England. Once home, Speke and Grant were welcomed in tumultuous fashion, although challenges from BURTON and others regarding the accuracy of Speke's geography produced considerable controversy. The matter occasioned widespread debate in newspapers, learned circles, and among the general public. Eventually, it was agreed that Burton and Speke would debate on the subject before a meeting of the British Association for the Advancement of Science, with the noted missionary/explorer, Dr. David LIVINGSTONE, acting as moderator.

The debate was scheduled for September 16, 1864, but it never took place. As a packed house listened in shocked silence, it was announced that Speke had shot himself while out hunting partridges the previous afternoon. The dramatic nature of his death added to the controversy surrounding the entire matter of the Nile's source, and the subject would be one of contention and uncertainty in Africanist and geographical circles for another generation. The nature of Speke's death, which was officially ruled accidental but which many believed, with some justice, was suicide, simply added a further element of poignancy. Only in 1890, after the travels of Baker, Henry Morton STANLEY, and others had added to knowledge of the Nile's headwaters, was the problem completely solved. In the end, Speke's claim to precedence as discoverer of the Nile's source was fully vindicated.

Impact

John Hanning Speke's fame rests entirely on his African travels, and his discovery of Lake Victoria was, as another African explorer, Sir Harry H. Johnston, said, the greatest geographical discovery since North America. His work, along with that of LIVINGSTONE, BURTON, STANLEY, and other African explorers, excited great public interest in Africa and thereby eventually fueled the flames of imperialism. In a sense, African discovery provided the foundation upon which the scramble for colonial possessions on the continent was built, and Speke did much to create the widespread interest in all things African which characterized the mid-Victorian era.

There can be no doubt that Speke was a contentious individual, but his single-minded mania for discovery loomed large in his success as an explorer. His legacy is one of having solved an age-old mystery and in so doing directing European attention to Africa in an unprecedented fashion. He remains in many ways an elusive figure, thanks to a paucity of surviving personal papers and a rather secretive nature, but recent research suggests that his was a complex personality, as was indeed generally the case with African explorers.

Bibliography

Bridges, Roy C. "Negotiating a Way to the Nile." In *Africa and Its Explorers*, edited by Robert I. Rotberg. Cambridge, Mass.: Harvard University Press, 1970. A useful overview of Speke's explorations set against the wider scope of nineteenth century African discovery.

Maitland, Alexander. *Speke*. London: Constable, 1971. The only full-length biography of Speke yet written, this work is flawed by inadequate research and an overemphasis on psychoanalysis. Nevertheless, it is the best available account of Speke's life.

Speke, John Hanning. *Journal of the Discovery of the Source of the Nile*. Edinburgh: William Blackwood, 1863. Speke's personal account of his final African journey.

James A. Casada

HENRY MORTON STANLEY

Born: January 28, 1841; Denbigh, Denbighshire, Wales
Died: May 10, 1904; London, England

Best known for finding and resupplying Dr. David Livingstone in 1871, Stanley was the first white man to chart a number of the great lakes in central Africa and follow the Congo River to its mouth. His exploration opened much of Africa to European commerce and colonization.

Early Life

Henry Morton Stanley, the illegitimate child of nineteen-year-old Elizabeth Perry, was born in Denbigh, Denbighshire, Wales, on January 28, 1841. He was christened John Rowlands, thus receiving the same name as his probable father, a twenty-six-year-old farmer. At the age of four, he entered the local grammar school, but two years later he was placed in St. Asaph Union Workhouse, where he received most of his formal education.

Stanley applied himself diligently to his studies, exhibiting a trait that would endure throughout his life. Though he was a man of action who could act impetuously, his successes in Africa owed much to the care with which he prepared his expeditions. A journalist who visited Stanley's apartment in 1874 noted that "the chairs, tables, sofas and settees, nay even the very floor itself—are laden with books, newspapers, manuscripts and maps" over which Stanley pored until late every night, making careful notes about weather, topography, and needed supplies.

Stanley also demonstrated his penchant for rashness at St. Asaph, for, according to his autobiography (1909), he ran away at the age of fifteen after knocking his teacher unconscious. He fled first to a cousin, Moses Owen, for whom he worked briefly as a teacher while receiving the last of his own classroom instruction. Owen's mother disliked her nephew, so after nine months Stanley had to leave his post as student-tutor. Instead, he earned a precarious living as a shepherd and bartender at Treneirchion, Wales, before crossing the border to Liverpool. From this port, in December, 1858, he sailed as a cabin boy aboard the *Windermere*, bound for New Orleans.

Once in the United States, Stanley again ran away from a desperate situation, jumping ship even though he was a stranger without prospects for employment. Luck favored him, though, and he was adopted by a generous cotton broker, Henry Hope Stanley, whose name he took; the "Morton" was a later addition. After clerking in New Orleans and Cypress Bend, Arkansas, Stanley enlisted in the Arkansas Greys when the Civil War began. Captured by Union forces at the Battle of Shiloh in April, 1862, he shifted his allegiance rather than languish in a prison camp. This phase of his military career ended quickly when he received a medical discharge less than a month after joining an Illinois regiment. He would make one further attempt to serve, signing with the federal navy on July 19, 1864. After seeing action off the coast of North Carolina, he jumped ship on February 10, 1865; Stanley was apparently born to command but not to serve.

A perpetual victim of wanderlust, in June, 1865, Stanley agreed to work as a free-lance writer for the Missouri *Democrat* (in St. Louis) to report on the Colorado gold rush. Over the next two years, this love of travel would carry him from the American West to the Middle East and back to his native Denbigh before he returned to Missouri.

Life's Work
Shortly after returning to the United States, Stanley seized an opportunity to travel once more. England was preparing to invade Abyssinia, and Stanley offered to cover the story for the New York *Herald*. Its owner, James Gordon Bennett, agreed—provided that Stanley would pay all of his own expenses. Stanley accepted these terms. Through bribery, skill, and luck, he was the first to report the British victory at the Battle of Magdala, thus securing a permanent spot on the staff of Bennett's paper.

Two years later, Bennett selected his young European correspondent for the mission that would make Stanley's name a household word and forever change the map of Africa. Dr. David LIVINGSTONE, missionary and geographer, had first gone to Africa in 1841. In 1866, he set out to find the source of the Nile River but soon disappeared. As concern for his safety mounted, Bennett asked Stanley to find and resupply the missing explorer.

Bennett's concern, though, was more journalistic than humanitarian; he recognized what Stanley's story would do for the circulation of the *Herald*. To be sure that the newspaper would have good copy regardless of the outcome of the rescue effort, Bennett first sent Stanley to cover the opening of the Suez Canal, on November 17, 1869, and to file a series of reports on the Middle East. Thus, Stanley could not begin his search for Livingstone until early 1871. On November 10, 1871, after overcoming bad weather, disease, and hostile tribesmen, Stanley found the man he was seeking, greeting him with the words that would soon echo throughout Europe and America: "Dr. Livingstone, I presume?" Stanley had given the *Herald* its story, and he had saved Livingstone, whose supplies were almost exhausted and who was suffering from dysentery. Livingstone tried to persuade Stanley to join him in his quest for the source of the Nile, while Stanley argued that Livingstone should return to England. Neither convinced the other, but Stanley was clearly captivated by the missionary and by Africa. Even though he was eager to tell his story to the Western world, Stanley tarried at Ujiji on Lake Tanganyika until March 14, 1872.

Back in London, Stanley described his life as a "whirl of cabs, soirées, dinners, dress-clothes and gloves." He was added to Madame Tussaud's waxworks; Queen Victoria sent him a jeweled snuffbox bearing the inscription, "Presented by Her Majesty, Queen Victoria to Henry Morton Stanley, Esq. in recognition of the prudence and zeal displayed by him in opening communication with Doctor Livingstone and thus relieving the general anxiety felt in regard to the fate of that distinguished Traveller." She also granted him an audience at Dunrobin Castle.

Not everyone was equally enthusiastic over Stanley's achievement, particularly since he pointed out that neither the Royal

Geographical Society nor the British government had done much to help LIVINGSTONE. Stanley was accused of fabricating the whole account and of forging the Livingstone letters and diary he brought back with him. In August, 1872, he bitterly described the treatment he had received from the Royal Geographical Society and the British upper classes:

> First they would sneer at the fact of an American having gone to Central Africa—then they sneered at the idea of his being successful. . . . My story is called "sensational" and unreal etc. I assure you that I think after decently burying Livingstone in forgetfulness they hate to be told he is yet alive.

When Stanley offered to accompany the belatedly organized relief expedition under the auspices of the Royal Geographical Society, he was rebuffed. Nevertheless, the society could not ignore the general sentiment concerning Stanley's success, and on October 31, 1872, it reluctantly awarded him its highest honor, the Victoria Gold Medal.

In the midst of speeches, dinners, honors, and receptions, Stanley found time to write two books about his recent adventures, *How I Found Livingstone* (1872) and *My Kalulu* (1873). The latter, a fictionalized adventure tale for youngsters, did not do well, but *How I Found Livingstone* was an immense success, selling more than seventy thousand copies.

Finding and resupplying LIVINGSTONE had captured the popular imagination; as Sir Clement Robert Markham observed, however, "The fellow has done no geography." The same could not be said of Stanley's next expedition. On November 17, 1874, he plunged into the African jungle again, determined to cross the continent from east to west, locate the source of the Nile, and chart the unknown equatorial regions.

Such a feat offered great rewards and equally large challenges. The weather could be oppressively hot, and torrential rains could quickly turn a campsite into a sea of mud. Under such miserable conditions, the natives who carried the supplies often would desert or mutiny. Dysentery, malaria, and typhus took a heavy toll: Stanley was the only white man in the party to survive the journey. Hostile tribes posed yet another danger; in a single battle at Vinyata, twenty-one people were killed by Nyaturu warriors. Within three months, Stanley had lost more than half of his original caravan.

The jungle exacted a price from Stanley himself also. In 1874 in London, he had weighed 178 pounds; since he was only five feet, five inches tall, his figure was portly. Yet he carried himself with military erectness, and photographs show arching brows above clear, gray eyes, a firm mouth, and flowing mustache. By the time he reached Ujiji on Lake Tanganyika early in 1876, he weighed only 118 pounds, and his black, wavy hair was streaked with gray.

Already, though, he had become the first white man to navigate and chart Lake Victoria, and he had discovered Lake George. Still greater achievements lay ahead. From Lake Tanganyika, he headed north along the Lualaba River, taking a route no European had ever followed and one that even natives feared, for the river was lined with cannibals. By choosing to go north rather than south, Stanley was to "do" geography indeed, for he not only discovered Stanley Falls and Stanley Pool, but also was to demonstrate conclusively that the Lualaba was part of the Congo rather than the Nile. Moreover, he would become the first European to travel the length of the Congo, for waterfalls and rapids blocked the way from the west, and previously the jungle had denied access from the east.

Again Stanley's feats aroused the interest of the Western world. Gold medals flowed in from learned societies in Europe and the United States; both houses of Congress gave Stanley a unanimous vote of thanks. Léon Gambetta of France summarized the popular sentiment of the day when he declared, "Not only, sir, have you opened up a new continent to our view, but you have given an impulse to scientific and philanthropic enterprise which will have a material effect on the progress of the world."

Such, in fact, were Stanley's aims. He hoped to end the slave trade in central Africa, Christianize the natives, and improve their material lot through commerce with the West. Events were to prove these plans illusory, for Stanley's explorations were to open the western half of the continent to Arab slave traders and equally exploitive European empire builders.

Stanley was still largely unaware of this perversion of his dreams when he agreed to return to the Congo for the Comité d'Études du

Haut Congo, organized by King Leopold II of Belgium. He would have preferred British support, but England was not interested. When Lieutenant Verney Lovett Cameron attempted to annex part of the Congo for the British Empire, Parliament refused the offer. Working for Leopold, Stanley earned his nickname, Bula Matari, "smasher of stones." Between 1879 and 1882, he surpassed the king's expectations, founding Leopoldville, building a three-hundred-mile wagon road from the Atlantic to Stanley Pool, and creating a small fleet to navigate the upper Congo River beyond. Because of these efforts, Belgium would receive 900,000 square miles in the center of the continent.

In November, 1886, Stanley began a lecture tour in the United States to recount his recent adventures, about which he had also written a book, *The Congo and the Founding of Its Free State* (1885). The work was sufficiently popular to be translated into seventeen languages. He was quickly recalled for his final trip to Africa, though, this time to rescue Emin Pasha (born Eduard Carl Schmitzev), a German posing as a Turk. This mysterious figure had been appointed governor of the Equatorial Province in Southern Sudan and had remained loyal to England after the followers of Muhammed Ahmed (the Mahdi) had overrun Khartoum and killed General Charles George Gordon, the British governor (1885).

For some reason, Stanley chose to approach Emin Pasha from the west. If he expected that the roads, stations, and fleet he had set up along the Congo would speed his journey, he was mistaken, for much of what he had established was already reverting to jungle. The march was in many ways a disaster. More than half the members of the expedition died along the way, and rumors of Stanley's death also circulated. On December 21, 1887, however, Parliament learned that Stanley had accomplished his mission of resupplying Emin, and the House of Commons rose in a body to give him an ovation in absentia.

Of all Stanley's forays into Africa, this last had the least enduring significance. Emin was better supplied than Stanley with everything but ammunition, and even this relief was not sufficient to prevent his being forced to flee shortly afterward. Yet Stanley was again the hero of the day. Already in Zanzibar, Stanley found piles of telegrams from world leaders. He had promised his publisher a book about the expedition, so he stayed in Cairo because he knew that in Europe he would be kept too busy to complete it. Meanwhile, in Berlin audiences watched a play about his adventures in Africa. In Reykjavík, people sang "Aurora Borealis" about his feats. Mugs and plates bearing Stanley's picture appeared in London shops, as did song sheets with titles such as "Stanley's Rescue" and "The Victor's Return." A lecture tour in the United States brought Stanley sixty thousand dollars; a similar round of talks in England added another ten thousand dollars.

The workhouse boy from Wales had become rich. He also finally found a wife. Twice before, he had been jilted, and in 1886, Dorothy Tennant had rejected his proposal. Now she changed her mind, and on July 12, 1890, they were married in Westminster Abbey. The ceremony assumed the scale of a coronation. The Abbey was filled with the prominent and the powerful, including William Ewart Gladstone, the Lord Chancellor, and the Speaker of the House of Commons. So popular was the event that five thousand people had to be turned away from the overcrowded church.

At his wife's urging, in 1892 Stanley gave up his American citizenship, which he had taken in 1885 to protect his royalties from piracy. Shortly afterward, he stood for Parliament; though he lost his first bid, he succeeded in 1895 and served for five years. To a man of action, the House of Commons held little fascination, however, and he complained of its "asphyxiating atmosphere." Honors continued to pour in; in 1899, he was awarded the Grand Cross of the Bath, thus becoming Sir Henry Morton Stanley, and the Atheneum, the most prestigious club in London, elected him a member.

Stanley still longed for action and Africa, but his four expeditions had aged him. In 1900, he retired to Surrey, where his wife named a local stream the Congo and a small lake Stanley Pool. These were the closest he would come to the places that had made him famous. He died in London on May 10, 1904.

Impact

Henry Morton Stanley had hoped to be buried in Westminster Abbey near LIVINGSTONE, with whom his name had been so closely linked. Although his funeral did occur there, he was denied burial in the historical shrine because by 1904 his name had also become associated with the Congo Free State, a synonym for atrocities.

"I was sent for a special work," Stanley had written in his autobiography, a work he had defined as "the redemption of the splendid central basin of the continent by sound and legitimate commerce." Through his explorations and his writings, he had opened up that area to reveal its richness to the Western world. His call for missionaries in the 1870's had brought such a response from England that Uganda became a British colony. He not only drew up the borders of the Congo Free State, later to become the Belgian Congo, but also ensured British involvement in the Sudan and eastern Africa through his various expeditions. As a journalist and explorer, he never failed to accomplish the missions on which he was sent. Whether he thereby effected Africa's redemption or damnation, though, remains an open question more than a century after he shed the first rays of light on Africa's heart of darkness.

Bibliography

Anstruther, Ian. *Dr. Livingstone, I Presume?* New York: E. P. Dutton, 1957. A detailed account of Stanley's early years and his rescue of Dr. Livingstone. Though the book touches briefly on Stanley's later achievements, it essentially ends in 1874.

Bierman, John. *Dark Safari: The Life Behind the Legend of Henry Morton Stanley.* New York: Alfred A. Knopf, 1990. Includes illustrations, bibliographical references, and an index.

Farwell, Byron. *Man Who Presumed: A Biography of Henry M. Stanley.* New York: Holt, Rinehart and Winston, 1957. A good biography for the general reader. Farwell adds no new information, but his account is clearly written and entertaining.

Hall, Richard. *Stanley: An Adventurer Explored.* Boston: Houghton Mifflin Co., 1975. A detailed biography, revealing many previously unknown details about Stanley's early life. Maps and photographs supplement the well-written text.

Hamshee, C. E. "Stanley's Second African Journey." *History Today* 18 (October, 1968): 713-721. An account of the organization, problems, and accomplishments of Stanley's journey across Africa from 1874 to 1877.

McLynn, Frank J. *Stanley*. Volume 1, *The Making of an African Explorer*, 1989. Volume 2, *Sorcerer's Apprentice*, 1991. London: Constable, 1989-1991. McLynn's volumes include illustrations, bibliographical references, and indexes.

Severin, Timothy. "The Making of an American Lion." *American Heritage* 25 (February, 1974): 4-11, 82-85. Stanley was passing himself off as an American two decades before he assumed United States citizenship. Severin tells Stanley's story from the American perspective, concentrating on his early life in the United States and the ecstatic American response to his achievements.

Smith, Ian R. *The Emin Pasha Relief Expedition, 1886-1890*. Oxford: Oxford University Press, 1972. A scholarly examination of Stanley's last African adventure. Draws on previously unpublished material to place the expedition in the context of European imperialism. Also recounts the harrowing experiences of both the Advance and Rear Columns trying to reach Emin Pasha.

Tames, Richard Lawrence Ames. *Henry Morton Stanley*. Aylesbury, England: Shire Publications, 1973. In less than fifty pages, Tames provides a good overview of Stanley's career. The brief text is richly illustrated to reveal both the man and his milieu.

Joseph Rosenblum

AMERIGO VESPUCCI

Born: March 14, 1454; Florence
Died: February 22, 1512; Seville, Spain

The first European credited with persuading his contemporaries that what Christopher Columbus had discovered was a "New World," Vespucci revolutionized geographic thinking when he argued that this region now bearing his name was a continent distinct from Asia.

Early Life

Amerigo Vespucci was the third son of a Florentine family of five children. His father, Stagio Vespucci, was a modestly prosperous notary and a member of a respected and learned clan that cultivated good relations with Florence's intellectual and artistic elite. The fortunes of the family improved during Amerigo's lifetime, and his father would twice occupy positions of fiscal responsibility in the Florentine government.

Unlike his older brothers, who attended the University of Pisa, Amerigo received his education at home under the tutelage of a paternal uncle, Giorgio Antonio, a Dominican friar. The youth became proficient in Latin and developed an interest in mathematics and geography, an interest which he was able to indulge in his tutor's extensive library. In his uncle's circle, Amerigo also became acquainted with the theories of Paolo Toscanelli dal Pozzo, a Florentine physician and cosmographer who first suggested the possibility of a westward voyage as an alternative route to the Orient, an idea that COLUMBUS and others eventually borrowed.

The study of geography was considered useful for anyone interested in a career in commerce, the profession chosen for Amerigo by his parents. Travel was also considered suitable training for businessmen, and Amerigo accepted the first opportunity when another

Library of Congress

uncle, Guido Antonio Vespucci, a lawyer, invited the twenty-four-year-old to Paris. The elder Vespucci had been appointed Florentine ambassador to the court of Louix XI in 1478 and asked his young relative to join him as his private secretary.

In 1482, two years after Amerigo's return to Florence from France, his father died, making Amerigo responsible for the support of the family. The following year, Amerigo became manager of the household of one of the branches of the ruling Medici family, and he performed his task loyally for the next sixteen years. In this capacity, he traveled to Spain at least once to look after the financial interests of the Medicis. He was in Spain again toward the end of 1491 and settled permanently in the city of Seville, where he established financial relations with the city's active Italian merchant community. He would eventually marry María Cerezo, a native of Seville. The couple had no children.

At the close of the fifteenth century, the port city of Seville was the hub of commercial activity and the center of overseas travel and exploration. The Portuguese had taken the lead in the search for a new route to India by reaching the Orient circumnavigating Africa. Confirmation of the accuracy of their vision came with news that Bartolomeu DIAS' expedition had reached the Cape of Good Hope (the southernmost tip of Africa) in 1488. The Spanish lagged behind their Portuguese neighbors until COLUMBUS' triumphant return from his first voyage. The Crown had paid Columbus' expenses, and he was expected to search for yet another alternate route to the East. Following the theories of Toscanelli, Columbus sailed in 1492 and returned to Spain early the following year.

Columbus' initial optimistic reports that he had found a new route to Asia ensured greater interest and opportunities for investment on the part of all who knew of his trip, and Vespucci would soon be involved in several of the many maritime enterprises that mushroomed in Seville in the wake of Columbus' success. Vespucci, as a subaltern of the Italian merchant Giannetto Berardi, assisted Columbus in financing and outfitting a second voyage of discovery, which sailed in 1493. Berardi died before the provisioning of the fleet was complete, and Vespucci assumed the task. It is highly likely that Vespucci and Columbus had many opportunities to meet during this period and that the Florentine's early interest in geography and cosmography was revived as a result of these contacts. The lure of the sea and the prospects of discovery would soon prove irresistible. By 1499, Vespucci had decided to change professions from businessman to explorer.

Life's Work
Much controversy surrounds certain facts about Vespucci's life between the years 1497 and 1499—the period immediately prior to his first generally acknowledged ocean voyage—especially because some of his biographers assert that he, not COLUMBUS, was the first European to discover the American mainland along the coast of northern South America. In order for this assertion to be valid, Vespucci would have had to undertake this voyage before Columbus' third—during which Columbus sailed along the coast of Venezuela—that is, before June, 1498. Vespucci was an inveterate letter writer. The most compelling evidence that he might have gone on this trip appears in a document of dubious authenticity attributed to Vespucci himself, the *Lettera di Amerigo Vespucci delle isole nouvamente trovate in quattro suoi viaggi* (c. 1505; *The First Four Voyages of Amerigo Vespucci,* 1885). This long letter is addressed to the head of the Florentine republic, the gonfalonier Piero Soderini. In this document, the author purports to have made four voyages overseas, the first of which, circa 1497, took him along the Caribbean coast of the American mainland—that is, to Venezuela, Central America, the Yucatán Peninsula, and the Gulf of Mexico, well in advance of Columbus. Since there is little independent evidence to corroborate information about this voyage, many scholars dismiss this episode as a fiction propagated by the letter, which could have been a forgery published by an overzealous and unscrupulous printer eager to cash in on a reading public thirsty for news of and reports from the New World. The fourth voyage described in the letter is also believed to be apocryphal.

What is universally accepted is the fact that Vespucci sailed for the New World as a member of a three-ship expedition under the command of the Spaniard Alonso de Ojeda in the spring of 1499. Two of the ships had been outfitted by Vespucci, at his own expense, in the hope of reaching India. Vespucci's expectations were founded on a set of maps drawn from the calculations of Ptolemy, the Egyptian mathematician and astronomer of the second century, whose work *Geographike hyphegesis* (*Geography*) was the foremost authority to fifteenth century Europeans on matters related to the size and shape of the world.

Ptolemy had concluded that the world was made up of three continents: Europe, Africa, and Asia. When Vespucci set out on his voyage in 1499, he expected to reach the Cape of Cattigara, the southernmost point of Asia on Ptolemy's map. Instead, his expedition reached the northern coast of Brazil and the mouth of the Amazon River. From there, Vespucci's ship proceeded southward to the equatorial zone, after which it turned northward to the Caribbean, navigating along the northeastern coast of South America. Seeing houses on stilts that reminded the crew of Venice, they named the area "Venezuela" (little Venice). The entire expedition returned to Spain, with a cargo of pearls and slaves but not the hoped-for Asian spices.

Back in Seville, Vespucci planned a second expedition that would take him farther south along the Brazilian coastal route, but his license to travel was suddenly revoked, on the grounds that he was a foreigner, when the Spanish crown, in competition with the Portuguese, began to treat geographical knowledge as secrets of state. When the ships that made up the expedition sailed in August, 1500, they carried only Spaniards. A Portuguese explorer, Pedro Álvars Cabral, had already claimed Brazil for the Portuguese crown in 1500 and, perhaps because of this fact, Vespucci's knowledge of its northern coast might have been of interest to Portugal. He was summoned to appear before King Manuel I. The monarch commissioned the Florentine to undertake a new voyage of discovery along the coast of Brazil, following Cabral's and Vespucci's own original intentions. Vespucci sailed from Lisbon in the spring of 1501.

This second independently verifiable voyage of Vespucci followed the coast of Brazil, crossed the equator, and proceeded south to Patagonia. Experiences during this last stage convinced Vespucci that Ptolemy's calculations had been mistaken, that the Cape of Cattigara and Asia were not where they were expected to be, and that the landmass before his eyes was more likely a new continent, separate and distinct from Asia. Upon his return to Lisbon, Vespucci, along with geographers and mapmakers, began to redraw and redesign Ptolemy's world to accommodate this new insight. The Atlantic coast of this region began to be detailed in maps that circulated throughout Europe, the first of which appeared in 1502.

Vespucci's employment by the Portuguese did not last long. He returned to Seville in 1502, disappointed that his plans for the exploitation of the new lands were not accepted by Manuel. In Spain, Vespucci's efforts and considerable geographical and navigational knowledge were finally recognized, and in 1505 he was granted citizenship by King Ferdinand II, who appointed him pilot major of the country's board of trade, the Casa de Contratación de las Indias. Vespucci held this position until his death in 1512.

Vespucci is believed to have been short of stature, with an aquiline nose, brown eyes, and wavy hair. This description comes from a family portrait painted by the Florentine muralist Ghirlandajo. Vespucci has also been described as deceitful, self-promoting, and cunning. His reputation suffered after the publication of two letters attributed to him, *The First Four Voyages of Amerigo Vespucci*, mentioned earlier, and *Mundus Novus* (c. 1503; English translation, 1916), an account of Vespucci's 1501 expedition addressed to Lorenzo de' Medici, his Florentine employer. In this second letter, the author argues that the lands he had recently visited (the Atlantic coast of South America) could only be part of a new world.

The ideas contained in the disputed letters, published in many editions and languages shortly after their initial printing, inspired a German mapmaker, Martin Waldseemüller, at Saint-Dié in Lorraine, to draw a new map to accompany narrative descriptions of this new world. The map, which was published in 1507, more closely resembles the geography of the American continent than earlier efforts, separates America from Asia, and assigns to the new land the name America in honor of its presumed discoverer Americus (Amerigo). The feminine version of Amerigo was selected to be consistent with the feminine names of the other continents, Europe, Africa, and Asia. This is the first known example of the use of America as the name of the new continent. The word was quickly accepted by northern Europeans as the rightful name for South America, but it would take some fifty years before southern Europe adopted the name and applied it to the entire American landmass, north and south.

Vespucci's complicity in this matter has never been fully established; some believe that he contributed to his own mythology by making himself the center of attention in all of his correspondence, never mentioning others in his circle under whose direction he might have worked. He is accused of taking credit for the deeds of his collaborators. Defenders of COLUMBUS, the bulk of Vespucci's critics, argue that the new continent should have been named for Columbus rather than for Vespucci the impostor. Columbus, however, was never quite convinced that the lands he had reached were not in Asia and did not live long enough to experience the historical slight in favor of Vespucci.

Impact

Amerigo Vespucci, in spite of the fact that he has been seriously criticized by a number of eminent and revered figures, deserves much of the credit for revolutionizing geographic thinking in Europe. His travels, especially his vain search for Asia following a Ptolemaic map, convinced him that the accepted authority on things geographical was mistaken. To challenge Ptolemy and a scientific tradition of such long standing in sixteenth century Europe was an act of great intellectual and moral courage. While Europeans were slow in accepting the full implications of Vespucci's discoveries, his insights nevertheless received much immediate publicity. Vespucci's ideas captivated the imagination of cartographers and publishers, and a steady stream of historical literature filled the minds of Europe's growing reading public. These accounts fired readers' imaginations. Vespucci's conclusions stimulated the growing community of cartographers, navigators, and geographers. He described his experiences in detail, kept careful records of astronomical, navigational, and geographical observations, and made it possible for his contemporaries to accept the idea of America long before additional eyewitness evidence would confirm the wisdom of his insights.

Bibliography

Arciniegas, Germán. *Amerigo and the New World: The Life and Times of Amerigo Vespucci*. New York: Alfred A. Knopf, 1955. A most admiring biography, which argues vehemently in favor of the authenticity of Vespucci's four voyages. The author dismisses some of the criticism of Vespucci as nationalistic propaganda.

Parry, J. H. *The Discovery of South America*. New York: Taplinger, 1979. An informative and panoramic account of European expansion in America by one of North America's most respected historians. This work is filled with replicas of contemporary maps and charts and is a serious and objective treatment of the period. Parry disputes the authenticity of *The First Four Voyages of Amerigo Vespucci* but credits Vespucci with having contributed to Europe's knowledge of geography and navigation.

Pohl, Frederick J. *Amerigo Vespucci, Pilot Major*. 2d ed. New York: Octagon Books, 1966. The author devotes much attention to Vespucci's mature years, the period of his life that coincides with his voyages overseas. Pohl believes that Vespucci was a most deserving individual and that his fame was legitimately earned. Contains a complete English version of two of Vespucci's letters and two informative appendices.

Vigneras, Louis-André. *The Discovery of South America and the Andalusian Voyages*. Chicago: University of Chicago Press, 1976. A carefully constructed survey of the separate expeditions from Spain to America beginning with Columbus' first voyage in 1492. A separate appendix is devoted to Vespucci's Portuguese voyage. The author's treatment of Vespucci echoes the consensus of contemporary scholarship about him by doubting the authenticity of two of the four voyages.

Zweig, Stefan. *Amerigo: A Comedy of Errors in History*. Translated by Andrew St. James. New York: Viking Press, 1942. An account by the popular Austrian writer who at one point resided in Brazil. Zweig believes that America received its name because of an error, and he argues that Vespucci's letters are filled with serious factual mistakes and coincidences. For Zweig, Vespucci's great fame rests on a false foundation.

Clara Estow

CHARLES WILKES

Born: April 3, 1798; New York, New York
Died: February 8, 1877; Washington, D.C.

Wilkes's determination and leadership as commander of the United States Exploring Expedition of 1838-1842 ensured the success of this major step in the emergence of the United States as a naval and scientific power.

Early Life

Born April 3, 1798, in New York City, Charles Wilkes was the youngest child of John Deponthieu Wilkes, who had immigrated to the United States from Great Britain during the Revolutionary War, and Mary Seton, whose father was her husband's business partner. Because of John D. Wilkes's success in business, the family was relatively affluent.

Wilkes's life was disrupted before his third birthday by the death of his mother. Thereafter, he was reared by various female relatives and friends, including his aunt, Elizabeth Seton (the first American saint, canonized in 1974). Subsequently educated at boarding and preparatory schools, Wilkes rejected Columbia College or a career in business for a life in the Navy. After three cruises on merchant ships during the years 1815-1817, Wilkes obtained his commission as a midshipman in the United States Navy in 1818.

His decision to pursue a naval career despite the objections of his father may have been an early indication of two of Wilkes's most enduring characteristics: his determination to do things his way and his self-assuredness. Once he had decided that a particular course of action was correct, he never wavered, no matter who criticized or objected. Add to this a bluntness in word and manner and a self-righteous attitude, and the result was a man in constant conflict with superiors, peers, and subordinates.

Life's Work

Wilkes served on a number of ships during the early years of his career. There were also extensive stretches of land duty and periods of leave while awaiting orders. The time during which he was not at sea was spent studying mathematics and the naval sciences: hydrography, geodesy, and astronomy. He studied with Ferdinand Hassler, the first superintendent of the United States Coast Survey, and gained experience in surveying. (During one of these periods on land, in April, 1826, just prior to his promotion to lieutenant, Wilkes married Jane Renwick, whom he had known since childhood.) In February, 1833, he became superintendent of the Depot of Charts and Instruments (the forerunner of the Navy Hydrographic Office and the Naval Observatory). By the mid-1830's, it was clear that he was one of the leading scientific minds in the United States Navy.

It was his scientific reputation, relatively minor among civilian scientists but outstanding compared to his fellow naval officers, which led to Wilkes's orders in March, 1838, to command the United States Exploring Expedition, despite his junior rank (he was thirty-ninth of forty lieutenants). It was not Wilkes's first contact with the Expedition, which would occupy more than two decades of his life. When the Expedition was first authorized in 1828, Wilkes had volunteered for duty with it; the idea of such an expedition was shelved a year later. In 1836, it was again authorized, and Wilkes was selected to travel to Europe to purchase the necessary scientific apparatus. When he returned in January, 1837, he found the Expedition in disarray and still far from ready for sailing. After rejecting sub-

Archive Photos

ordinate positions with the Expedition, he accepted the responsibility as commander.

On August 18, 1838, the United States Exploring Expedition, usually known as the Wilkes Expedition, set sail from Norfolk, Virginia. Among its objectives were the surveying and charting of portions of the Pacific Ocean for the benefit of the American shipping industry, the establishment of good relations with the native populations of the region, and scientific observations and collecting in a number of disciplines. There were six ships, five hundred sailors, and nine civilian scientists under the command of the clean-shaven lieutenant whose hair always appeared disobedient. In a typical act of audacity, Wilkes assumed the rank of acting captain upon leaving port, despite the fact that the secretary of the Navy had denied him that appointment.

The Expedition returned to New York in the summer of 1842, having circumnavigated the earth and cruised more than eighty-seven thousand miles. Only two of the original ships survived the entire cruise; two had sunk, one was sent back early, and one was sold, while another ship had been added in 1841 to replace one of the lost ships. The Expedition had explored, surveyed, charted, and mapped the Pacific Ocean basin from Oregon to Australia. Its exploration of the coast of Antarctica confirmed that the great ice mass was in fact a continent. More than four thousand zoological, fifty thousand botanical, and thousands of ethnographical specimens were brought back by the explorers. Most of these scientific specimens were eventually deposited in the Smithsonian Institution in 1858.

The first order of business, however, were courts-martial. Wilkes brought charges against a number of his junior officers, and in turn, a number were brought against him. He was found guilty on only one charge, excessive punishment of sailors, and sentenced to a public reprimand. Characteristically, in his autobiography (1979), Wilkes dismisses his court-martial as a vendetta on the part of the secretary of the Navy.

With the courts-martial out of the way, Wilkes focused his energy on ensuring that the fruits of the Expedition would not be lost. In the summer of 1843, he was promoted to commander and given the responsibility for the Expedition's collections and reports. He published the five-volume *Narrative of the United States Exploring Expedition* in 1844, taking credit for authorship, although in fact he functioned more as an editor. He lobbied Congress for the funding for, and oversaw the publication of, nineteen scientific reports, authoring those on meteorology and hydrography himself. In 1855, his rank was raised to captain. Except for survey duty in 1858, the Expedition remained his only official concern until the outbreak of the Civil War.

Wilkes's personal life changed during these years. In 1848, his wife, Jane, with whom he had four children, died. Six years later, he married Mary Lynch Bolton, with whom he had two additional children, although one, born when Wilkes was sixty-seven, did not survive infancy.

During the Civil War, Wilkes held a number of successful commands and was promoted to commodore. There was, however, controversy during this time as well. While captain of the *San Jacinto* in November, 1861, he removed two Confederate agents from the British mail packet *Trent*. Although he was hailed as a hero by the American public and government, members of the latter changed their attitude when it was realized that Wilkes's action could precipitate a war with Great Britain. The agents were freed. A subsequent clash with Secretary of the Navy Gideon Welles led to a second court-martial of Wilkes in 1864 for disobedience of orders, a finding of guilty, and suspension from the Navy.

In 1866, Wilkes was promoted to rear admiral on the retired list. He continued working for the publication of the final volumes of the Expedition reports, although Congress ceased their funding in 1873 with a number of reports unpublished. Wilkes died in his home in Washington on February 8, 1877. In 1909, his remains were placed in Arlington National Cemetery.

Impact

In the 1830's, American science was just beginning to step upon the international stage. Although scientific exploration, a common enough activity among the great European powers, had been attempted with some success by the United States Army, the United

States Exploring Expedition represented an effort on a scale far larger than the republic had ever attempted. The possibility of failure was high.

Despite the odds, the Expedition succeeded. The Expedition met its military objectives, and its scientific achievements placed the United States in a position of intellectual respectability. A model was provided for subsequent naval exploratory expeditions. To a very large extent, credit for that success must go to Wilkes. He was a difficult man to get along with, could not tolerate opposition or criticism, and knew far less about the world, especially science, than he thought he did. Yet, when energy, drive, and determination were needed, whether aboard ship or in the halls of Congress, Wilkes met the challenge. James Dwight Dana, the young geologist on the Expedition, thought that no other naval officer could have done better. Combining a dedication to duty with a sympathy for science, Wilkes left the world a legacy of scientific and geographical knowledge.

Bibliography

Henderson, Daniel M. *The Hidden Coasts: A Biography of Admiral Charles Wilkes*. New York: William Sloane Associates, 1953. The first full-length biography of Wilkes, this is an example of the pro-Wilkes partisan literature. It handles the scientific activities of the Wilkes Expedition very poorly and is of questionable accuracy.

Ponko, Vincent, Jr. *Ships, Seas, and Scientists: U.S. Naval Exploration and Discovery in the Nineteenth Century*. Annapolis: Naval Institute Press, 1974. Surveys the naval exploring expeditions of the antebellum period. Places the Wilkes Expedition in its larger context.

Smith, Geoffrey Sutton. "The Navy Before Darwinism: Science, Exploration, and Diplomacy in Antebellum America." *American Quarterly* 28 (1976): 41-55. Treats antebellum naval exploration as part of the diplomatic efforts of the United States to secure a commercial empire.

Stanton, William. *The Great United States Exploring Expedition of 1838-1842*. Berkeley: University of California Press, 1975. Provides an analysis of the scientific contributions of the Wilkes Expedition. Stanton argues that the Expedition gained international respect for the American scientific community.

Tyler, David B. *The Wilkes Expedition: The First United States Exploring Expedition, 1838-1842*. Philadelphia: American Philosophical Society, 1968. Concentrates on the exploring and surveying aspects of the Expedition rather than the scientific. This history focuses on the performance of Wilkes as a commander and his relationship with his officer corps and civilian scientists.

Viola, Herman J., and Carolyn Margolis, eds. *Magnificent Voyagers: The U.S. Exploring Expedition, 1838-1842*. Washington, D.C.: Smithsonian Institution Press, 1985. A collection of articles by scientists and historians. This represents the latest and most balanced account of Wilkes and his expedition.

Wilkes, Charles. *Autobiography of Rear Admiral Charles Wilkes, U.S. Navy: 1798-1877*. Edited by William James Morgan, David B. Tyler, Joye L. Leonhart, and Mary F. Loughlin. Washington, D.C.: Naval History Division, 1979. A massive defense of his career, this autobiography provides unique insight into the mind of the man. Both the positive and the negative aspects of his character shine through.

Marc Rothenberg

Sir George Hubert Wilkins

Born: October 31, 1888; East Mount Bryan,
South Australia, Australia
Died: December 1, 1958; Framingham, Massachusetts

Wilkins was able to utilize new technological developments and to apply aviation, cinematography, and meteorology in order to understand the diverse conditions of the polar regions during his explorations.

Early Life

George Hubert Wilkins was born at East Mount Bryan, near Adelaide, South Australia, on October 31, 1888. He was the thirteenth and youngest child of Harry Wilkins and the former Louisa Smith. Harry Wilkins had failed to find his fortune in the Ballarat, Victoria, gold strikes in 1851 and turned to the open range as one of the earliest drovers to bring cattle into South Australia. It was on his sheep and cattle ranch that Wilkins lived and worked as a young boy. Although he received a diploma qualifying him to enter a state high school, Wilkins had no formal secondary education because he spent nearly three years helping his father through a devastating drought. His years of living in the vastness of the country, observing and camping with the neighboring aborigines and experiencing the destructive forces of nature, influenced Wilkins' lifetime interest in natural sciences, anthropology, climatology, and meteorology.

In 1903, his parents retired to Adelaide, where Wilkins worked in the mornings and attended classes at both the University of Adelaide and the South Australian School of Mines and Industries. Although he studied electrical and general engineering, his interests diversified to include music, botany, zoology, geology, and particularly photography. While attending school, Wilkins served as an apprentice to a mechanical engineer and later spent nearly a year in charge of the electric lighting for a touring carnival company. It was during his years with the carnival that he developed his love for travel and his expertise in the new technology of motion pictures.

In 1908, Wilkins was offered a position with the Gaumont Motion Picture Company in London, England, as a cinematographic cameraman. He stowed away on a ship, was caught and forced to work on the ship's dynamo, and eventually arrived in England after an adventurous journey through the Mediterranean and North Africa. In London, his rare skills with cameras and motion pictures enabled him to work for both the Gaumont company and the London *Daily Chronicle*. While on assignment at Hendon Aerodrome, Wilkins met Claude Grahame-White, the pioneer English aviator, who took him on his first flight and arranged for him to take flying lessons. Wilkins' flight training was interrupted when his employers jointly sent him to cover the brutal Balkan War in 1912-1913 as a cinematographer. He was briefly captured during the war and at great personal risk became the first photographer to obtain motion pictures of actual combat. After the war, Wilkins continued to build a reputation in the photographic and cinematographic fields in Europe and the West Indies while continuing his flying lessons in both airplanes and dirigibles. In 1913, his career suddenly changed when he was invited to join an expedition to the Arctic.

Life's Work
In 1913, Wilkins joined the Vilhjalmur Stefansson Arctic Expedition, sponsored by the Canadian government. During the following three years, he walked thousands of miles across the Arctic ice and acquired great expertise in the techniques of living, traveling, and working under Arctic conditions. Although Wilkins was the expedition's photographer, he added greatly to his knowledge of natural sciences, studied Eskimo ethnology, and carried out oceanographic and meteorological experiments. He also developed plans to utilize airplanes in polar explorations and mapping and for establishing permanent weather stations in those regions as part of a worldwide weather forecasting program.

Because of World War I, Wilkins left the expedition in 1916 to accept a commission in the Royal Australian Flying Corps. He was

Library of Congress

assigned to the military history department as a photographer, navigator, and pilot. During this assignment, he was wounded nine times and received the Military Cross with Bar for his bravery. When the war ended, Wilkins returned to flying. In 1919, he participated in the England-to-Australia air race, but a fuel leak forced his Blackburn Kangaroo airplane to land in Crete.

In 1920-1921, Wilkins made his first trip to Antarctica as second in command of John Lachlan Cope's British Imperial Expedition to survey the coastline of Graham Land. On his second visit, he served as naturalist with Sir Ernest Shackleton and John Quiller Rowett's *Quest* expedition of 1921-1922, during which Shackleton died. Wilkins discovered several new species of vegetation, birds, and insects, while continuing his meteorological study of the polar regions.

Wilkins spent 1922 and 1923 in Europe and the Soviet Union with the Society of Friends, filming the effects of drought and famine. Upon his return to London, he was selected by the British Museum to lead a natural history expedition in 1923-1925 to collect plant and animal specimens in tropical Northern Australia. He summarized his results in his book *Undiscovered Australia* (1928), which showed the extent and quality of his studies of plants, birds, insects, fish, mammals, fossils, and archaeological artifacts.

Unable to secure funding for an Antarctic expedition, Wilkins, with support from several private sources, such as the *Detroit News* and the American Geographical Society, returned to his earlier goal of exploring the Arctic by airplane. After two abortive efforts in 1926 and 1927, Wilkins and Carl Ben Eielson in 1928 flew their Lockheed Vega monoplane twenty-one hundred miles across previously unexplored territory between Point Barrow, Alaska, and Spitsbergen, Norway. In his book *Flying the Arctic* (1928), Wilkins explained that the flight was made to prove the value of airplanes for polar exploration and to further his plans for polar meteorological stations.

Wilkins was knighted by King George V on June 14, 1928, at Buckingham Palace, in recognition of his pioneer flights and other accomplishments. He won several other honors, including the Patron's Medal of the Royal Geographic Society and the Samuel Finley Breese Morse Gold Medal of the American Geographical Society.

Wilkins next launched his project for the aerial exploration of the Antarctic as the leader of the Wilkins-Hearst Expedition sponsored by the American Geographical Society and financier William Randolph Hearst. From Deception Bay in the South Shetland Islands, he and Eielson made the first flight in the Antarctic on November 16, 1928. On subsequent flights from 1928 to 1930, Wilkins flew over Graham Land, discovering Crane Channel, Stefansson Strait, and the Lockheed Mountains. He claimed the island known as Charcot Land for Great Britain and mapped more than eighty thousand square miles of Antarctica.

Wilkins' reputation as a pioneer aviator in polar regions was firmly established. Handsome, six feet tall, and possessing a grace that belied his solid two hundred pounds, Wilkins was highly respected for his scholarship, professionalism, amiability, and integrity. In 1929, he married Suzanne Bennett, an Australian actress who lived in New York. Theirs was a childless but happy marriage which by mutual agreement permitted them to pursue their individual careers.

In all of his flights over the Arctic Sea, Wilkins had found no land on which to build his projected weather stations. Unwilling to accept ice floes as station platforms and doubting that surface ships could penetrate the icy seas to suitable locations, he turned to the submarine as an experimental weather station. Converting an obsolete navy submarine, which he named the *Nautilus*, into an oceanographic laboratory, Wilkins proposed crossing the Arctic basin using the vessel both above and below the ice to radio weather information to the world. A series of mishaps and malfunctions forced the *Nautilus* expedition of 1931 to be abandoned, but not before the feasibility of using submarines under the polar ice cap had been demonstrated. This was the first submarine trip under the Arctic ice and preceded the atomic-powered *Nautilus* by twenty-seven years.

From 1933 to 1937, Wilkins commanded the ship *Wyatt Earp* and managed four Antarctic flights by the American Lincoln Ellsworth. In November, 1935, Ellsworth, utilizing Wilkins' planning, succeeded in the first flight across the Antarctic continent, rendezvousing with Wilkins aboard the *Wyatt Earp*. In August, 1937, the Soviet Union gave Wilkins the command of a search expedition to locate Soviet aviator Sigesmund Levanevsky, who had disappeared between Moscow and Alaska. During the following months, Wilkins and Herbert Hollick-Kenyon combined the search with pioneer moonlight flying in winter conditions while covering more than

150,000 square miles of the uncharted polar basin, but Levanevsky was never found.

The Levanevsky search was Wilkins' last great feat of exploration. During World War II, Wilkins was utilized as a geographer, climatologist, and Arctic adviser to the United States Quartermaster Corps and the Office of Strategic Studies. He was specifically involved in developing clothing and equipment for troops engaged in rugged environments. Wilkins designed special parkas and underwear for troops assigned to polar regions and personally tested them in the Aleutian Islands.

After the war, Wilkins worked with the United States Navy Office of Scientific Research from 1946 to 1947 and served as an adviser to the United States Weather Bureau (1946-1948) and the Arctic Institute of North America (1947). He was a guest lecturer in geography at McGill University in Montreal, Canada, in 1947-1948 and at the National Defense College of Canada in 1948.

Wilkins was an active participant in the Antarctic studies conducted during the International Geophysical Year, 1957-1958. Wilkins died of a heart attack on December 1, 1958, in Framingham, Massachusetts, where he had worked as a consultant on polar regions for the Research and Development Command of the Department of Defense since 1953. On March 17, 1959, the American nuclear submarine *Skate*, the first vessel to surface at the exact geographical North Pole, honored Wilkins' lifetime wish by scattering his ashes over the icy terrain.

Impact

Sir George Hubert Wilkins was primarily a field explorer who had an inquisitive mind and the outlook of a true pioneer. He was at home with primitive people whose reverence for nature he shared. His adaptability to extreme environmental conditions made him an ideal polar explorer. At the same time, he was one of the least publicly recognized expedition leaders because he cared less for headlines than for genuine advances in knowledge of humankind's environment. He was never interested in the races to either pole, and it was only after thousands had preceded him that he visited either one, even though he had spent five summers and twenty-six winters in the Arctic and eight summers in the Antarctic.

In addition to his pioneering efforts in cinematography and the *Nautilus* expedition, Wilkins was the first to fly in the Antarctic, the first to fly over the Arctic Ocean, and the first to prove the feasibility of landing a plane on packed ice. Respect from his colleagues and honors from governments were bestowed on him for his feats and scientific contributions. His greatest successes and rewards were the reorientation of geographic thought that the airplane engendered, the development of submarines capable of exploring the polar waters, and the establishment of the polar weather stations for which he had worked during his lifetime.

Wilkins was as much at home with the scientific world of the International Geophysical Year in 1957 as he had been with his early experiences in cinematography and flight. His ability to utilize new technology and ideas and to realize their long-term significance made him the consummate explorer that he was.

Bibliography

Bertrand, Kenneth J. *Americans in Antarctica, 1775-1948*. New York: Lane Press, 1971. An American Geographical Society special publication. Excellent information about Wilkins with particular emphasis on the Ellsworth flights.

Grierson, John. *Sir Hubert Wilkins: Enigma of Exploration*. London: Robert Hale, 1960. A factual and informative biography of an admirable individual whose adventurous life was lived in near obscurity.

Kirwan, Laurence P. *A History of Polar Exploration*. New York: W. W. Norton and Co., 1960. Comprehensive narrative with references to Wilkins in the context of polar exploration.

MacLean, John Kennedy, and Chelsea Fraser. *Heroes of the Farthest North and Farthest South*. New York: Thomas Y. Crowell, 1932. Good chapters on the early expeditions and accomplishments of Wilkins and Stefansson.

Mill, Hugh Robert. "The Significance of Sir Hubert Wilkins' Antarctic Flights." *The Geographical Review* 19 (July, 1929): 377-386. A contemporary and professional analysis of the importance of Wilkins' early aeronautical achievements in the Arctic.

Stefansson, Vilhjalmur. *Unsolved Mysteries of the Arctic*. New York: Macmillan, 1938. An entire chapter is devoted to Wilkins' involvement in the Levanevsky search. Comprehensive narrative by an expert in polar history.

Thomas, Lowell. *Sir Hubert Wilkins, His World of Adventure*. New York: McGraw-Hill Book Co., 1961. Informative biography in which the author has Wilkins narrate his life from their thirty years of conversations. Excellent photographs.

Wood, Walter A. "George Hubert Wilkins." *The Geographical Review* 49 (July, 1959): 411-416. Brief but informative biographical account of Wilkins' life and accomplishments.

Phillip E. Koerper

WILLIAM OF RUBROUCK

Born: c. 1215; Rubrouck, French Flanders
Died: c. 1295; place unknown

William provided the first accurate account of the geography of Central Asia and of its people, the Mongols. He thus helped to fill in a blank space on the map and opened up a new era of exploration.

Early Life

The life of William of Rubrouck (Willem van Ruysbroeck) remains shrouded in mystery despite scholarly efforts to shed light on it. He appears on the stage of history between May, 1253, and August, 1255, the period during which he undertook a journey to the court of the Great Khan in Mongolia. Except for a brief stay in Paris during the late 1250's or early 1260's, nothing is known about his life before or after his historic journey.

Scholars assume that William was from the village of Rubrouck in Flanders, the northeasternmost corner of modern France. The date of his birth is unknown, though some historians place it as early as 1215. Similarly, the year of his death is unknown, though it is assumed that he was still alive when Marco POLO returned from his journey to China in 1295.

Nothing is known of William's educational background. When Louis IX commissioned him to go to Mongolia, William was a Franciscan friar serving at the king's court. The saintly Louis IX was fond of the mendicant (begging) orders of monks and so surrounded himself with friars. William's own narrative of his journey provides the only insight into his learning and character. Though the work is not written in the best Ciceronian Latin, the author reveals himself as a keen observer, one who was able to sift the relevant from the irrelevant and thus provide Europe with its first truly reliable information about the geography and peoples of inner Asia.

William reveals himself as a bold, even daring adventurer. The brazenness with which he preached the Christian faith shocked his Mongolian hosts, whose religious tolerance no doubt puzzled and angered William. He records in his narrative that several times the Great Khan urged him to be more diplomatic in his debates with Muslims, Buddhists, shamanists, and Nestorians. Fear that William was disrupting the religious peace may have been one reason that the Great Khan ordered him to return to his home.

Life's Work

It was only his journey to Mongolia that lifted William of Rubrouck from obscurity onto the pages of history. Hence, a discussion of his life's work must focus upon a period of roughly twenty-eight months. Beyond that, he is all but unknown.

William and Louis IX had similar yet different motives for a mission to Mongolia (Tartary). William's motives were primarily religious; Louis' motives were a mixture of religious and political. William was apparently deeply moved by reports of the plight of German slaves of the khan of the Golden Horde (Russia and Kazakhstan). The Germans were Catholics, and William felt burdened to go and minister the sacraments to them. He was encouraged by rumors that some of the Mongol rulers had already accepted, or were on the verge of accepting, Christianity.

Louis IX, noted for his piety, was similarly influenced by the rumors that certain of the khans were Christians. He was also encouraged, however, by the prospects of an alliance with the Mongols against the Muslims in the Middle East. Louis hoped to learn something about the intentions of the Mongol armies in Syria. Thus, he was persuaded to overcome his reluctance to send William. William was not to travel as an accredited envoy of the French king, however, but as a Christian missionary seeking permission to settle, found a mission, and preach the Gospel among the Mongols. Louis gave William a letter addressed to Prince Sartach, an alleged Christian and eldest son of Batu Khan, ruler of the Golden Horde, requesting that William be given safe conduct and permission to preach.

William was commissioned by Louis IX in the spring of 1252,

while Louis was resident in Acre, Palestine, following a disastrous crusade in Egypt. From Acre, William journeyed to Constantinople, where he preached in the church of Hagia Sophia on Palm Sunday, 1253. On May 7, 1253, he departed from Constantinople on the first leg of his historic journey.

William's party consisted of four individuals. In addition to himself, there was a fellow Franciscan, Bartholomew of Cremona. Bartholomew was later to remain at the court of the Great Khan and become the first Catholic missionary to die in the East. There was also a clerk to look after the gifts Louis was sending and an interpreter, who proved unreliable.

William and his party reached Sartach's camp at Sarai, where the Volga River empties into the Caspian Sea, on July 31. During the three days he remained with Sartach, William learned that the Mongol prince was neither a Christian nor really interested in religious matters. Sartach ordered William to proceed to the court of his father, Batu Khan. Batu was encamped near Saratov, on the upper reaches of the Volga River, in modern eastern Russia.

Batu in turn sent William on to Mangu, the Great Khan himself, and provided two Nestorian Christian guides for the journey. They reached Mangu's encampment on the northeastern slopes of the Altai Mountains on December 27, 1253. They were treated courteously by Mangu, though he remained suspicious of William's true motives.

In the spring of 1254, Mangu returned to his capital at Karakorum, capital of the vast Mongol Empire. William and his party went with him and remained at Karakorum until July, when Mangu ordered him to return home. William was regularly questioned by Mangu's ministers, who apparently were never fully convinced that he was not an ambassador. Since he insisted that he was only a Christian missionary, he did not have the right to request an audience with the Great Khan; William had to wait for Mangu to summon him.

During his stay at Karakorum, William was housed with Nestorian monks. From time to time, Mangu himself arranged for disputations between William and representatives of the various religions of his subjects. Mangu was obviously very proud of the Mongols' tradition of religious toleration. William noted that the khan was careful not to show any preference for any one religion; he diplomatically spread his patronage equally among Buddhists, Taoists, shamanists, Muslims, and various Christian sects and attended all of their important ceremonies. William's dogmatic advocacy of his own Christian faith offended Mangu, apparently leading to Mangu's decision to send William back to Europe.

William departed Karakorum in July, 1254. He carried with him a letter from Mangu to Louis IX, calling upon the great lords and priests of Europe to go to Karakorum and do homage to the Great Khan. On June 16, 1255, he arrived in Cyprus, where he was disappointed to find that Louis IX had returned to France. Though he desired to go to Paris and report personally to Louis IX, he was ordered by his provincial vicar to return to Acre, where he became a lecturer in theology. This turn of events was fortuitous, for it forced William to write a narrative of his journey. All in all, William had traveled some ten thousand miles, much of it on horseback over harsh terrain.

Impact

William of Rubrouck's narrative of his journey provided a wealth of information for Europeans. His geographical revelations restored knowledge that had been lost to Western Europeans since the fall of the Roman Empire in the West (A.D. 476). He confirmed that the Caspian Sea was in fact an inland sea. He was the first European to recognize that Cathay was "Seres," the mythical city where ancient and medieval Westerners believed silk originated.

William was also the first European to describe an Asian city. Though he found Karakorum less impressive than Paris, what he described was a metropolitan capital of a vast, pluralistic empire. Not only did he note the existence in Karakorum of twelve Buddhist, Taoist, and shamanist temples, two Muslim mosques, and one Nestorian church, but he also provided the first descriptions of the religious rites and practices of those religions.

Karakorum was a meeting place of the various Asiatic peoples ruled over by the Mongols. William observed and recorded their lifestyles, folklore, and customs. He was the first person to make

Chinese writing known to Europeans. By his description of Karakorum and its varied residents, William dispelled the traditional belief that Asian cities contained palaces made of gold and precious gems. In the same way, his observations of Central Asia laid to rest the popular belief that the area was inhabited by mythical monsters.

After his return, William served as lecturer in theology at Acre. Perhaps as a result of the intervention of Louis IX, William was given permission by his vicar to return to Paris in the late 1250's or early 1260's. There he met the English scientist and philosopher Roger Bacon, a fellow Franciscan. Much of William's narrative was incorporated by Bacon in his *Opus majus* (1267; English translation, 1897-1900), which he acknowledges was written with William's help.

In a subsequent study, published in the mid-1260's, Bacon recorded the formula for gunpowder. Modern scholars believe that Bacon had obtained the formula from William, who in turn had learned of it while in Karakorum. Thus, William of Rubrouck's legacy was a mixed one. On the one hand, he opened up to Europeans a whole new world. On the other hand, he may well have given Europe gunpowder, thus helping usher in the era of modern warfare.

Bibliography

Chambers, James. *The Devil's Horsemen: The Mongol Invasion of Europe*. New York: Atheneum Publishers, 1979. Chambers provides a highly readable survey of the Mongolian invasion of Europe. Several clear maps help the reader locate geographical and battle sites mentioned in the text. Chapter 10 includes a summary of William of Rubrouck's journey, putting it in its historical context.

Dawson, Christopher, ed. *The Mongol Mission: Narratives and Letters of the Franciscan Missionaries in Mongolia and China in the Thirteenth and Fourteenth Centuries*. New York: Sheed and Ward, 1955. Part of the Makers of Christendom series. In addition to providing the text of William of Rubrouck's narrative in a very readable translation, Dawson's lengthy introduction includes all that is known about William and his mission. William's journey is placed in historical context with the other thirteenth century journeys of exploration to Mongolia.

Olschki, Leonardo. *Marco Polo's Asia: An Introduction to His "Description of the World" Called "Il Milione."* Translated by John A. Scott. Berkeley: University of California Press, 1960. Though this study focuses on Marco Polo, Olschki also deals with Polo's predecessors, including William of Rubrouck. It is a scholarly but readable discussion of the European discovery of Asia.

Prawdin, Michael. *The Mongol Empire: Its Rise and Legacy.* Translated by Eden Paul and Cedar Paul. London: Allen and Unwin, 1961. Chapter 18 is especially helpful for its discussion of the Great Khan's religious tolerance.

Rockhill, William Woodville, ed. and trans. *The Journey of William of Rubruck to the Eastern Parts of the World, 1253-55, with Two Accounts of the Earlier Journey of John of Pian de Carpine.* London: Hakluyt Society, 1900. Reprint. Nedeln, Liechtenstein: Kraus Reprint Limited, 1967. This is the most authoritative English translation of William's narrative. The text is accompanied by rich explanatory footnotes, maps, and an itinerary of William's journey. A thirty-two-page introduction provides an excellent summary of Europe's relations with the Mongols in the thirteenth century, as well as the background to William's journey.

Paul R. Waibel

JOHN WINTHROP

Born: January 12, 1588; Edwardstone, Suffolk, England
Died: March 26, 1649; Boston, Massachusetts

Winthrop was committed to the ideal of creating a Christian commonwealth, and his determined leadership was crucial to the establishment of the Massachusetts Bay Colony.

Early Life

John Winthrop was born January 12, 1588, in the small English village of Edwardstone in Suffolk. His mother, née Anne Browne, and his father, Adam Winthrop, lived at Groton Manor in Suffolk. Winthrop's grandfather, also named Adam, a successful London cloth merchant, had purchased the manor from Henry VIII in 1544. It had been part of a monastery confiscated by the monarch. Winthrop received an extensive education beginning at age seven with instruction from a local vicar. At fifteen, he entered Trinity College, Cambridge. Winthrop remained there less than two years, but later studied law at Gray's Inn, one of the London Inns of Court. He started a family at a tender age; his father ar-

Archive Photos

ranged a marriage to Mary Forth when he was only seventeen, and he became a father at eighteen. Over the next twelve years, Winthrop moved from his dowry lands at Great Stambridge back to Groton, presided over the manorial court, served as a justice of the peace, and assumed control of the family lands on the manor. In the 1620's, he widened his horizons by developing a lucrative London law practice. It enabled Winthrop to make contacts in the government and led to his selection in 1627 as an attorney in the King's Court of Wards and Liveries, a court which administered the estates of minor heirs to lands held from the king.

As he matured, Winthrop became ever more committed to the faith of the Puritan reformers in the Church of England. Advocates of the teaching of John Calvin, Puritans believed that God had predestined salvation for only a few. From his early teens, Winthrop had followed a rigorous regimen of prayer and study in search of signs that God had selected him. While he struggled for assurance of that elect status, Winthrop also sought to place his relationship with God ahead of all else—his family, his work, and his love of hunting, food, and drink. He never became an ascetic; he believed God's creations should be enjoyed but always with proper moderation. His faith made Winthrop a stern and determined man; this was clearly evident in a painting of the mature Winthrop. His serious countenance, graced by a Vandyke beard and ruffled collar, befits a man with a sense of purpose.

Life's Work
Winthrop probably would have been known only as one of the lesser English gentry had not a series of economic, religious, and personal crises in the late 1620's caused him to leave England. Inflation, smaller returns from the fixed rents he could charge the tenants on his land, and a depression in the Suffolk textile trade all dearly cost the squire of Groton. His disappointing financial situation worsened in 1629 when Winthrop lost his attorneyship in the Court of Wards and Liveries. He was only one of numerous casualties in the campaign of Charles I to remove Puritans from secular and religious positions. The king's Catholic wife, the appointment of William Laud (a resolute anti-Puritan) as Bishop of London, and the dissolution in

1629 of a Parliament heavily influenced by Puritans all caused dissenters to despair about their future in England. They did not see much hope for their faith under a monarch who opposed their advocacy of simpler services and a Calvinist theology in the Church of England. Winthrop had a more immediate reason for feeling that he was living in an evil and declining nation. Long concerned by what he considered a lax moral climate in the nation, Winthrop was appalled by the behavior of his son Henry. The nineteen-year-old had gone to the West Indian island of Barbados in 1627. Upon his return two years later, Henry did little more than carouse with boisterous friends in London. Winthrop believed that it was imperative that he act to save his family and preserve his faith. He worried that migration meant abandoning his homeland, but he hoped that the creation of a model Christian community in North America would show England the way to reform.

Winthrop worked with the members of the Massachusetts Bay Company to achieve his goal. Made up of substantial landowners, merchants, and clergymen, the company selected Winthrop as its governor. He organized the ships, settlers, and provisions for the expedition and then led more than one thousand people to Massachusetts in 1630. In the next nineteen years, Winthrop retained an important role in the colony's government; twelve times he was elected governor. Throughout those years, he struggled to keep the settlers committed to building a cooperative, godly commonwealth.

Challenges to Winthrop's vision emerged quickly. Few were willing to settle in a single, compact town as Winthrop had hoped; in addition to Boston, six towns were formed in the first year alone. The cheap land, the high wages paid to scarce skilled workers, and the profits to be made in commerce led many to a greater concern over the material benefits of Winthrop's colony than its spiritual. Price and wage controls mitigated the impact of the more acquisitive settlers but could not completely suppress the growing economic individualism. More troubling to Winthrop than the greed of some colonists and the dispersal of settlement, however, was religious dissent. Winthrop and his supporters did not migrate to Massachusetts to create a utopia of toleration; rather they moved to worship in a singular fashion—in self-governing congregations of

God's elect. Consequently, Winthrop fought all attempts to undermine that effort.

The first significant trouble came from Roger Williams, a minister who arrived in the colony in 1631. Among other things, Williams demanded that the colonists repudiate all ties to the Church of England, a position contrary to Winthrop's resolve to reform, not break from, the established church. Williams also argued that the elect should not worship with the unregenerate, an idea repugnant to Winthrop, whose hope for a unified colony dictated that persons of all conditions worship together. A brilliant woman, Anne Hutchinson, presented an even greater threat. Not long after her arrival in 1634, she began to hold mid-week meetings in her home. Hutchinson used these popular gatherings to criticize ministers whom she believed erred in their sermons. All but two, she charged, taught a Covenant of Works—that good conduct could lead to salvation—rather than a Covenant of Grace—that salvation was obtained only through God's grace. Her attacks on the clergy made Hutchinson a danger to the established order. When she later claimed that she received divine revelation, Hutchinson became a pariah to Winthrop. When persuasion failed to convince the two dissenters to retreat from their positions, Winthrop supported the decision to banish them, Williams in 1636 and Hutchinson two years later. Other dissenters also suffered banishment or were forced to migrate to other regions. The departure of these people ensured a religious orthodoxy that prevailed throughout the colony's first two decades.

In addition to these religious conflicts, Massachusetts faced other vexing problems. Disputes over who could participate in the government were resolved by permitting male church members to vote. Complaints from outlying settlements that their interests were not being served by Boston lawmakers were handled by allowing each town representation in the colony's General Court. The outbreak of civil war in England in the early 1640's dramatically reduced the immigration to Massachusetts. Immigrants had been the chief consumers of the colony's produce, and with the decline in their numbers, the economy slumped badly. Prices fell until the Puritans found new markets in the Canaries and the Caribbean Islands.

Winthrop figured prominently in the resolution of these difficulties; he helped work out the political problems, and he maintained contacts and promoted trade in the West Indies.

Winthrop usually avoided the extremes in both secular and religious matters. He deplored the ideas of separatist dissenters such as Williams, for example, because he believed that they would lead to the chaos of dozens of little utopias. Yet other leaders in the colony, notably Thomas Dudley, criticized Winthrop for being too lenient with dissenters. His general commitment to moderate positions offended many, but it helped preserve the Puritan experiment in the New World, one with more than fifteen thousand inhabitants at his death in 1649. Besides a grateful colony, Winthrop was survived by six of his sixteen children and his fourth wife.

Summary

As he led the Puritan expedition across the Atlantic in early 1630, Winthrop had time to think about the meaning of their collective effort. He drafted a lay sermon containing those reflections, and he delivered it to the passengers prior to their arrival in Massachusetts. Entitled "A Modell of Christian Charity," it remains one of the most eloquent statements of Christian brotherhood. He explained to his followers

> wee must be knitt together in this worke as one man, wee must entertaine each other in brotherly Affeccion . . . wee must delight in eache other, make others Condicions our owne rejoyce together, mourne together, labour, and suffer together, allwayes haveing before our eyes . . . our Community as members of the same body. . . .

His effort to convince settlers to subordinate their self-interest to the good of the community was far from successful. Yet through his example and his support of laws governing economic behavior, Winthrop helped keep in check the individualism he believed would destroy the colony.

Perhaps Winthrop's greatest impact on American life was his evocation of a sense of mission. He thought that the tired generations of the Old World were eagerly observing the Puritan effort to build a model religious society. "For wee must Consider," he claimed, "that wee shall be as a Citty upon a Hill, the eies of all people are uppon us." Winthrop and fellow Puritans believed that they were God's new chosen people, a new Israel. Succeeding generations have shared this sense that America had a special destiny to be a light to other nations. They have revealed their debt to the great Puritan leader each time they borrowed his metaphor and claimed that America must be as a city upon a hill.

Bibliography

Bremer, Francis J. *The Puritan Experiment: New England Society from Bradford to Edwards*. New York: St. Martin's Press, 1976. Bremer provides a comprehensive account of the Puritans through the mid-eighteenth century. In addition, he includes a helpful discussion of secondary works and a guide to the most important primary sources on the Puritans.

Miller, Perry. "Errand into the Wilderness." In *Errand into the Wilderness*. Cambridge, Mass.: Harvard University Press, 1956. This is an essay by the leading historian on the Puritan mind. He describes both the exhilarating sense of mission Winthrop shared with other leaders in the 1630's and the disappointment of a later generation when it realized that England had paid scant attention to the errand of reform they had run for God.

Morgan, Edmund S, ed. *The Founding of Massachusetts: Historians and the Sources*. Indianapolis: Bobbs-Merrill Co., 1964. This is a helpful collection of primary sources from the first five years of the Massachusetts Bay Colony. Notably, Morgan includes more than one hundred pages from Winthrop's letters, journal, and miscellaneous other papers. There are also excerpts from the works of four historians' accounts of the colony.

_____. *The Puritan Dilemma: The Story of John Winthrop*. Glenview, Ill.: Scott, Foresman and Co., 1988. Morgan's book is not only the best available biography of Winthrop, but also one of the clearest presentations of Puritan thought. He discusses Winthrop's life in England and America and in the process details the struggle faced by a pious man in a corrupt world.

Morison, Samuel Eliot. *Builders of the Bay Colony*. Boston: Houghton Mifflin, 1930. Reprint, 1958. Morison profiles more than a dozen individuals in these lively essays originally published in 1930. The profiles attempt to rehabilitate the long-tarnished image of Puritans, and they serve as an excellent introduction to the leading personalities in seventeenth century Massachusetts. The longest is on Winthrop, and in it Morison portrays him as a pious yet practical leader.

Moseley, James G. *John Winthrop's World: History as a Story, the Story as History.* Madison: University of Wisconsin Press, 1992. Part of the History of American Thought and Culture series. Includes bibliography and index.

Rutman, Darrett B. *Winthrop's Boston: A Portrait of a Puritan Town, 1630-1649.* Chapel Hill: University of North Carolina Press, 1965. A well-researched and well-written work on the Puritan capital during Winthrop's life. A study of the town's government, church policies, population trends, and economic development, it reveals how far Bostonians strayed from Winthrop's goal of a cooperative godly community.

Schweninger, Lee. *John Winthrop.* Boston: Twayne, 1990. Part of Twayne's United States Authors series; includes a bibliography.

Vaughan, Alden T., and Francis J. Bremer, eds. *Puritan New England: Essays on Religion, Society and Culture.* New York: St. Martin's Press, 1977. This collection of essays, written by leading scholars, covers many areas of Puritan life—religion, witchcraft, government, economics, family, and race relations. Several include references to Winthrop.

Wall, Robert E. *Massachusetts Bay: The Crucial Decade, 1640-1650.* New Haven, Conn.: Yale University Press, 1972. A detailed account of the political events of the 1640's. Wall describes the growing conflict between leaders from Boston and those in other towns jealous of their power.

Larry Gragg

FANNY BULLOCK WORKMAN

Born: January 8, 1859; Worcester, Massachusetts
Died: January 22, 1925; Cannes, France

A tireless explorer and geographer, writer, accomplished linguist, feminist, and suffragist, Fanny Bullock Workman set international mountain-climbing records for women. Her enormous contribution to the body of geographical knowledge was acknowledged by numerous geographical societies around the world.

Early Life

Fanny Bullock was born into a wealthy family in Worcester, Massachusetts, on January 8, 1859. Her mother was Elvira Hazard Bullock. Fanny's maternal grandfather was Augustus George Hazard, a merchant and gunpowder manufacturer based in Connecticut, where he built up the family fortune. Fanny's father, Alexander Hamilton Bullock, was a politician who served as the Republican governor of Massachusetts from 1866 to 1868. Fanny had an older sister and brother. Her early education came from private tutors. After completing Miss Graham's Finishing School in New York, she spent two years in Dresden and Paris, where she became fluent in German and French. She returned to Massachusetts when she was twenty. At the age of twenty-two, on June 16, 1881, Fanny was married to William Hunter Workman, a physician. He was twelve years older than she was, had done his postgraduate studies in Munich, and had already traveled extensively in Europe. They had one daughter, Rachel, in 1884. Fanny began hiking with her husband in the White Mountains of New Hampshire. It was there that she climbed her first mountain, Mount Washington (6,293 feet), an unusual accomplishment for a woman of that time. In 1886, they began taking trips to Scandinavia and Germany. William Workman became ill in 1888, and since they were independently wealthy, he retired from his medical practice without causing them any economic hardship. The Workmans spent the next nine years in Europe, using Germany as their home base while they traveled, leaving their child in the care of nurses or at boarding school. It was during these years that Fanny did her first serious climbing.

Life's Work

Fanny Bullock Workman, who preferred to be called Mrs. Bullock Workman, began her adventurous career when her husband took her hiking in the White Mountains of New Hampshire. She climbed Mount Washington several times. After her husband retired and they moved to Europe, she began to make her first serious ascents.

Most of the climbing that Fanny did during their early years in Europe was in the Alps. With the help of guides, she scaled Zinal Rothorn (4,221 feet), the Matterhorn (14,780 feet), and Mont Blanc (15,781 feet). These were exceptional accomplishments, because it was unacceptable in the 1890's for women to do mountain climbing.

Amazingly, she made these climbs wearing the long skirts that were considered proper for women of that era. In fact, throughout the years of her exploring and climbing, she continued to wear skirts as a part of her outfit, though in later years she did begin to wear them shortened up to her boot tops.

Wearing skirts was Fanny Workman's only concession to the feminine role that was considered appropriate in the Victorian age. She and her husband were adamant in their belief in the equality of women with men. As their excursions grew longer and more complex, they began trading roles from year to year. One would organize the expedition, arranging for all the necessary supplies, pack animals, permits, workers, and guides. The other would be responsible for all the photography and record keeping. Both tasks were enormous. Their expedition parties grew to include more than a hundred people, and many arrangements had to be made long distance via mail and telegraph. The records that they kept during these expeditions included precise scientific readings of geographic location and altitude, mapping, and geological descriptions of the terrain. Hundreds of photos were taken with the best equipment then available—bulky, heavy cameras and tripods that had to be carried in cumbersome wooden cases.

In the early 1890's, Fanny and her husband began going on bicycle tours, first in Europe and then in North Africa. These journeys were not mere sightseeing trips; they were adventures. The Workmans faced attacks by wild dogs, journalists eager for interviews, bandits, extremes of weather, poor food and water supplies, epidemics of malaria and the plague, and other problems that would have stopped less determined travelers. They began writing collaborative accounts of their adventures, and the first book they published was *Algerian Memories: A Bicycle Tour over the Atlas Mountains to the Sahara* (1895). In that same year, they took with them the recently invented Kodak camera to the Iberian Peninsula. The book that followed was *Sketches Awheel in Modern Iberia* (1897). The book recording their longest journey, which they took from 1897 through 1899, was *Through Town and Jungle: Fourteen Thousand Miles Awheel Among the Temples and People of the Indian Plain* (1904). This trip also involved traveling 1,800 miles in Ceylon and 1,500 miles in Java, Sumatra, and Cochin China (South Vietnam). These books all had many good reviews and were well received by a wide audience.

The part of this longest journey that had the most impact on them was a side trip that they took to escape the intense summer heat while in India in 1898. In Kashmir, they put aside their bicycles for a few weeks and proceeded on foot to see the Karakoram and Himalayan mountain ranges up close. They were so enchanted that they put together an expedition the next year, planning to return to Sikkim to spend two months hiking and climbing there.

The venture in Sikkim was beset by problems from the beginning. The Workmans had never arranged such a major venture before, were unaccustomed to the terrain and the climate, and were unfamiliar with the local customs and language. It had taken so long to arrange the expedition that, by the time they got started, the weather—which had already been unseasonably bad for some weeks—was worsening with the approaching winter, and the days were growing short. They were determined, however, and they set off with their large caravan and staff in October. The couple's eagerness and spirit of adventure were not shared by the porters and bearers. These workers were used to less-determined mountaineers who did not insist on risking the arduous journey under such dangerous weather conditions or traveling at such a fast pace.

Despite their convictions regarding the equality of women, the Workmans treated their hired workers with astonishing insensitivity. In the Workmans' account of this expedition, *In the Ice World of the Himálaya* (1900), they showed that they had not risen above the American social model of the time—racism. Not recognizing the impact of their lack of experience and the environmental conditions, let alone the devastating effects of their leadership style, they placed the blame for the nearly overwhelming problems of this expedition on their perception that the Asian workers were uncooperative and unmanageable.

The Workmans never modified their approach when working with their porters and bearers in any of their further ventures in the Karakoram or Himalayan ranges, and they suffered many enormous hardships because of it. In one expedition in the Karakoram, 150 of their workers deserted, taking huge amounts of staple foods with them.

The work that Fanny and her husband did in their seven expeditions in the Himalayas and Karakoram ranges was remarkable and invaluable, and it included many firsts. Fanny set altitude records—as high as 23,000 feet—for women that went unmet for decades. They mapped uncharted areas, including some of the largest nonpolar glaciers in the world. Their observations were essential to geological knowledge of glacial processes. Their maps were the first records of the watersheds for several rivers in the areas bordering Nepal and Tibet. They wrote five books recording these expeditions—the one previously mentioned and *Ice-Bound Heights of the Mustagh* (1908), *Peaks and Glaciers of Nun Kun* (1909), *The Call of the Snowy Hispar* (1910), and *Two Summers in the Ice-wilds of Eastern Karakoram* (1917). They also wrote articles for magazines such as *The National Geographic* and *Alpine Journal*.

Fanny's professional recognition by scholars and boards of geographical societies came slowly. It was not an era when women were accepted as knowledgeable or capable of such undertakings. It was not only the sheer volume of precise data that she had collected but also the documentation of the care that had been taken to collect it that won them over. They may have been swayed also by the length of her career in such daunting expeditions. The peak recognition that she received was from the Royal Geographic Society, where she lectured in 1905, becoming only the second woman to have done so.

After World War I, the Workmans retired for good in the South of France. Fanny was ill for several years before she died at the age of sixty-six in Cannes, France.

Impact

Fanny Bullock Workman excelled as an explorer, climber, and geographer at a time when women were expected to be fragile and helpless. Her accomplishments were recognized by geographic societies and academic institutions around the world.

Because Bullock Workman spoke several languages, she could usually communicate directly with people in many of the places she traveled. She delivered lectures in several countries in their national language. She was the first American woman to speak at the Sorbonne.

Honors from ten European nations' geographical societies were bestowed on Fanny. She was a member of the Royal Asiatic Society and was a fellow of the Royal Geographical Society and the Royal Scottish Geographical Society. In the United States, she was a Corresponding Member of the National Geographic Society and the Brooklyn Institution of Arts and Science. She was a charter member of the American Alpine Club and an Honorary Member of the Appalachian Mountain Club.

Fanny was an ardent feminist. In 1912, she was photographed at an altitude of 21,000 feet on the Silver Throne plateau in the Himalayas, reading a newspaper. Its headline proclaims "Votes for Women." She believed strongly in higher education for women, and to that end she willed a total of $125,000 to Bryn Mawr, Radcliffe, Smith, and Wellesley, which were then exclusively women's colleges. She believed that women should be granted equal status with men in the scientific, social, literary, and political fields.

In her private life, she and her husband were patrons of the arts. They were great fans of the music of Richard Wagner, literature, and art. The two were devoted to each other, and their marriage was a partnership in both their personal and professional lives.

Bibliography

Hamalian, Leo, ed. *Ladies on the Loose: Women Travellers of the Eighteenth and Nineteenth Centuries*. New York: Dodd, Mead, 1981. The chapter on Fanny Bullock Workman in this book provides limited biographical information and then an excerpt from *Through Town and Jungle*, which is about bicycling in India. It is the only book that Bullock Workman wrote without her husband. Her comments regarding the native peoples are careful, detailed, and objective.

McHenry, Robert, ed. *Liberty's Women*. Springfield, Mass.: G. and C. Merriam, 1980. This volume includes brief but detailed biographical information. No photos or maps are included. No specific information on any specific expedition is given.

Miller, Luree. *On Top of the World: Five Women Explorers in Tibet*. New York: Paddington Press, 1976. A balanced, very readable account. Discusses some of the controversy that surrounded the couple's treatment of the hired workers and guides during the 1898 expedition in Sikkim. Includes studio photos of Fanny Bullock Workman.

Workman, Fanny Bullock, and William Hunter Workman. *In the Ice World of Himálaya: Among the Peaks and Passes of Ladakh, Nubra, Suru, and Baltistan*. New York: Cassell, 1900. Their first book about the Workmans' Himalayan expeditions. The narration is uneven in content, though it is interesting. In it are harsh comments about the workers they hired. Many photos and illustrations are provided. Includes a chapter in two parts, one by each author, detailing physiological responses to high altitudes.

_____. *Two Summers in the Ice-wilds of Eastern Karakoram: The Exploration of Nineteen Hundred Square Miles of Mountain and Glacier*. New York: E. P. Dutton, 1917. The body of this book was written by Fanny, which may account for its warm, personal tone. It includes fine geologic and geographic observations and detailed descriptions. There are also numerous photographs, many of which are fold-out panoramas, of the expedition in progress. Several scientific tables are included.

Marcella Joy

Time Line of Explorers

Date	Name	Nationality	Locale of Exploration	Endeavors Other than Exploration
c. 520 to 510 B.C.-?	Hanno	Carthaginian	West Africa	Colonization; Geography; Navigation
c. 350 to 325 B.C.- after 300 B.C.	Pytheas	Greek	North Atlantic	Astronomy; Scientific exploration; Geography; Navigation
c. 602-664	Hsüan-tsang	Chinese	Asia	Religion; Literature; Geography; Ethnography
c. 890-956	al-Mas'udi	Arab	Arabia; North Africa	Historiography; Geography; Literature
c. 970- c. 1035	Leif Eriksson	Norwegian	Northeastern North America	Mariner
c. 1180- 1252	Giovanni da Pian del Carpini	Italian	North Africa; Asia; Europe	Religion; Historiography
c. 1215- c. 1295	William of Rubrouck	French	Central Asia	Geography; Ethnography; Religion
c. 1254- 1324	Marco Polo	Italian	Asia	Geography; Ethnography
1304- c. 1377	Ibn Battutah	Arab	Asia; Russia; North Africa; East Indies	Ethnography; Literature

EXPLORERS

Date	Name	Nationality	Locale of Exploration	Endeavors Other than Exploration
c. 1371-between 1433 and 1436	Cheng Ho	Chinese	South Asia; Southeast Asia; East Africa	Government; Mariner; Navigation; Military
1394-1460	Prince Henry the Navigator	Portuguese	West Africa	Military; Government; Geography
c. 1447-after 1526	Pêro da Covilhã	Portuguese	India; Arabia; East Africa	Geography; Diplomacy; Military
c. 1450-1500	Bartolomeu Dias	Portuguese	Indian Ocean; Cape of Good Hope; South Atlantic	Mariner; Navigation
c. 1450-c. 1498	John Cabot	Italian	Eastern Canada; South America	Geography; Navigation
1451-1506	Christopher Columbus	Italian/ Spanish	Caribbean; Central America	Navigation; Mariner
1454-1512	Amerigo Vespucci	Italian	Caribbean; South America; Central America	Geography; Navigation
1460-1521	Juan Ponce de Léon	Spanish	Caribbean; Florida	Spanish conquest; Military; Colonization
c. 1460-1524	Vasco da Gama	Portuguese	India; South Atlantic; Indian Ocean; Asia	Military; Navigation

TIME LINE OF EXPLORERS

Date	Name	Nationality	Locale of Exploration	Endeavors Other than Exploration
1475-1519	Vasco Núñez de Balboa	Spanish	Caribbean; Central America; Pacific Ocean	Spanish conquest
1480-1521	Ferdinand Magellan	Portuguese/ Spanish	World circumnavigation; Pacific Ocean	Mariner; Navigation; Geography
1485-1547	Hernán Cortés	Spanish	Caribbean; Central America	Military; Spanish conquest
c. 1490- c. 1560	Álvar Nuñez Cabeza de Vaca	Spanish	American South; Mexico	Spanish conquest; Geography; Topography
c. 1491- 1557	Jacques Cartier	French	Northeastern Canada	Mariner; Navigation; Colonization
c. 1495- 1541	Francisco Pizarro	Spanish	Caribbean; Central America; South America	Spanish conquest; Colonization
c. 1496- 1542	Hernando de Soto	Spanish	Southeastern North America; Central America; South America	Spanish conquest; Military
1510-1554	Francisco Vásquez de Coronado	Spanish	American Southwest	Spanish conquest; Colonization; Government

583

EXPLORERS

Date	Name	Nationality	Locale of Exploration	Endeavors Other than Exploration
1519-1574	Pedro Menéndez de Avilés	Spanish	Florida; American Southeast	Spanish conquest; Military; Mariner; Colonization
c. 1535-1594	Martin Frobisher	English	Northwest Passage; Arctic; Subarctic; Canada	Mariner; Piracy; Military
c. 1540-1596	Sir Francis Drake	English	World circumnavigation	Navigation; Military; Piracy
c. 1550-1605	John Davis	English	Northwest Passage; Subarctic; South Atlantic	Mariner; Navigation; Piracy
1552 or 1554-1618	Sir Walter Ralegh	English	North America; South America	Government; Literature; Colonization; Piracy; Military; Navigation
1560-1592	Thomas Cavendish	English	World circumnavigation	Navigation; Piracy; Guide
1560's-1611	Henry Hudson	English	Northeastern North America; Northwest Passage	Navigation; Mariner
c. 1567-1635	Samuel de Champlain	French	Northeastern North America	Colonization; Geography

TIME LINE OF EXPLORERS

Date	Name	Nationality	*Locale of Exploration*	*Endeavors Other than Exploration*
1580-1631	John Smith	English	Eastern North America	Colonization; Geography; Government; Literature
1588-1649	John Winthrop	English/ American	Eastern North America	Government; Colonization; Religion
1643-1687	Sieur de La Salle	French	Central North America; Mississippi River	Colonization
1644-1718	William Penn	English/ American	Eastern North America	Religion; Government; Colonization
1651-1715	William Dampier	English	World circumnavigation; Pacific Ocean; Caribbean	Navigation; Mariner; Scientific exploration; Piracy
1681-1741	Vitus Jonassen Bering	Danish	Siberia; American Northwest; Arctic; Subarctic	Mariner; Navigation
1696-1785	James Edward Oglethorpe	English	Southeastern North America	Colonization; Military
1728-1779	Captain James Cook	English	World circumnavigation; Pacific Ocean	Oceanography; Scientific exploration; Field exploration; Mariner; Geography; Colonization; Navigation

EXPLORERS

Date	Name	Nationality	Locale of Exploration	Endeavors Other than Exploration
1729-1811	Louis-Antoine de Bougainville	French	World circumnavigation; South Pacific	Military; Colonization; Navigation
1730-1794	James Bruce	Scottish	African interior	Geography; Ethnography
1734-1820	Daniel Boone	American	American West	Frontiersman; Colonization
1743-1820	Sir Joseph Banks	English	World circumnavigation; Australia; New Zealand; Eastern Canada	Natural science; Botany; Scientific exploration; Colonization
c. 1764-1820	Sir Alexander Mackenzie	Scottish	Western Canada	Geography
1770-1838	William Clark	American	American Northwest	Military; Frontiersman; Geography; Diplomacy
1771-1806	Mungo Park	Scottish	West Africa; Niger River	Botany; Geography
1774-1809	Meriwether Lewis	American	American Northwest	Military; Scientific exploration
1779-1813	Zebulon Montgomery Pike	American	American Southwest	Military; Frontiersman; Topography; Geography
c. 1788-1812	Sacagawea	American	American Northwest	Guide; Geography; Diplomacy

TIME LINE OF EXPLORERS

Date	Name	Nationality	Locale of Exploration	Endeavors Other than Exploration
1793-1836	Stephen Fuller Austin	American	Texas; American Southwest	Colonization; Government; Statesmanship
1798-1877	Charles Wilkes	American	Antarctic; Pacific Ocean; World circumnavigation	Field exploration; Military; Mariner; Scientific exploration
1799-1831	Jedediah Strong Smith	American	American West	Frontiersman
1809-1868	Kit Carson	American	American West	Frontiersman; Military; Diplomacy; Guide
1813-1890	John C. Frémont	American	American West	Government; Military; Geography; Scientific exploration
1813-1873	David Livingstone	Scottish	Africa	Geography; Religion
1821-1890	Sir Richard Francis Burton	English	Asia; Africa; South America	Scholarship; Diplomacy; Ethnography; Archeology; Geography; Literature
1827-1864	John Hanning Speke	English	East Africa; Central Africa	Geography; Military

EXPLORERS

Date	Name	Nationality	Locale of Exploration	Endeavors Other than Exploration
1834-1902	John Wesley Powell	American	American West; Grand Canyon	Field exploration; Scientific exploration; Natural history; Topography; Geology; Geography; Ethnography
1838-1914	John Muir	Scottish/American	American West	Conservation; Literature; Scientific exploration
1841-1904	Henry Morton Stanley	British/American	Central Africa; West Africa	Geography; Literature; Government
1856-1920	Robert Edwin Peary	American	Arctic	Military
1859-1925	Fanny Bullock Workman	American	Europe; North America; Asia	Mountaineering; Geography; Field exploration
1861-1930	Fridtjof Nansen	Norwegian	Arctic; North Atlantic	Scientific exploration; Statesmanship; Humanitarianism; Zoology; Oceanography; Field exploration

TIME LINE OF EXPLORERS

Date	Name	Nationality	Locale of Exploration	Endeavors Other than Exploration
1872-1928?	Roald Amundsen	Norwegian	Northwest Passage; Arctic; Antarctic	Field exploration; Scientific exploration; Mariner
1879-1933	Knud Johan Victor Rasmussen	Danish	Arctic	Ethnography; Geography; Anthropology
1888-1957	Richard E. Byrd	American	Arctic; Antarctic	Aviation; Military
1888-1858	Sir George Hubert Wilkins	Australian	Arctic; Antarctic	Aviation; Field exploration; Natural science; Engineering; Scientific exploration; Anthropology; Military
1897-1937?	Amelia Earhart	American	American and transoceanic skies	Aviation
1902-1974	Charles A. Lindbergh	American	American and transoceanic skies	Aviation; Military; Engineering; Conservation
1910-	Jacques-Yves Cousteau	French	Undersea	Oceanography; Field exploration; Scientific exploration; Engineering; Biology; Zoology

EXPLORERS

Date	Name	Nationality	Locale of Exploration	Endeavors Other than Exploration
c. 1910-1980	Jacqueline Cochran	American	American and transatlantic skies	Aviation; Military
1914-	Thor Heyerdahl	Norwegian	Atlantic Ocean; Pacific Ocean; Indian Ocean	Mariner; Ethnography; Field exploration; Anthropology; Archeology; Scientific exploration; Navigation
1919-	Sir Edmund Hillary	New Zealander	Mount Everest; Antarctic	Mountaineering; Humanitarianism
1930-	Neil Armstrong	American	Space	Aviation; Aeronautics
1942-	Robert Duane Ballard	American	Undersea	Field exploration; Scientific exploration; Engineering; Archeology; Oceanography
1951-	Sally Ride	American	Space	Aviation; Aeronautics; Astrophysics; Scientific exploration

EXPLORERS

INDEX

A

Aeronautics
  Neil Armstrong, 10
  Sally Ride, 487
Africa
  Sir Richard Francis Burton, 74
  David Livingstone, 349
Africa (central)
  James Bruce, 68
  John Hanning Speke, 524
  Henry Morton Stanley, 530
Africa (eastern)
  Cheng Ho, 138
  Pêro da Covilhã, 196
  John Hanning Speke, 524
Africa (northern)
  Giovanni da Pian del Carpini, 101
  Ibn Battutah, 309
  al-Mas'udi, 372
Africa (western)
  Hanno, 262
  Prince Henry the Navigator, 269
  Mungo Park, 405
  Henry Morton Stanley, 530
American explorers
  Neil Armstrong, 10
  Stephen Fuller Austin, 19
  Robert Duane Ballard, 33
  Daniel Boone, 55
  Richard E. Byrd, 81
  Kit Carson, 108
  William Clark, 329
  Jacqueline Cochran, 145
  Amelia Earhart, 230
  John C. Frémont, 238
  Meriwether Lewis, 329, 337
  Charles A. Lindbergh, 338
  John Muir, 383
  Robert Edwin Peary, 412
  William Penn, 422
  Zebulon Montgomery Pike, 430
  John Wesley Powell, 457
  Sally Ride, 487
  Sacagawea, 494
  Jedediah Strong Smith, 502
  Henry Morton Stanley, 530
  Charles Wilkes, 547
  John Winthrop, 567
  Fanny Bullock Workman, 574
American Northwest
  Vitus Jonassen Bering, 48
  William Clark, 329
  Meriwether Lewis, 329
  Sacagawea, 494
American West
  Daniel Boone, 55
  Kit Carson, 108
  John C. Frémont, 238
  John Muir, 383
  John Wesley Powell, 457
  Jedediah Strong Smith, 502
Amundsen, Roald, 1-9
Antarctic
  Roald Amundsen, 1
  Richard E. Byrd, 81
  Sir Edmund Hillary, 284
  Charles Wilkes, 547
  Sir George Hubert Wilkins, 553
Anthropology
  Thor Heyerdahl, 277
  Knud Johan Victor Rasmussen, 480
  Sir George Hubert Wilkins, 553
Arab
  Ibn Battutah, 309
  al-Mas'udi, 372
Arabia
  Pêro da Covilhã, 196
  al-Mas'udi, 372

Archeology
  Robert Duane Ballard, 33
  Sir Richard Francis Burton, 74
  Thor Heyerdahl, 277
Arctic/Subarctic
  Roald Amundsen, 1
  Vitus Jonassen Bering, 48
  Richard E. Byrd, 81
  John Davis, 210
  Martin Frobisher, 248
  Fridtjof Nansen, 389
  Robert Edwin Peary, 412
  Knud Johan Victor Rasmussen, 480
  Sir George Hubert Wilkins, 553
Armstrong, Neil, 10-18
Asia
  Sir Richard Francis Burton, 74
  Giovanni da Pian del Carpini, 101
  Vasco da Gama, 255
  Hsüan-tsang, 292
  Ibn Battutah, 309
  Marco Polo, 443
  Fanny Bullock Workman, 574
Asia (central)
  William of Rubrouck, 561
Asia (southeastern)
  Cheng Ho, 138
Asia (southern)
  Cheng Ho, 138
Atlantic Ocean
  Thor Heyerdahl, 277
Austin, Stephen Fuller, 19-26
Australia
  Sir Joseph Banks, 39
  Captain James Cook, 161
Australian explorer
  Sir George Hubert Wilkins, 553
Aviation
  Neil Armstrong, 10
  Richard E. Byrd, 81
  Jacqueline Cochran, 145

Amelia Earhart, 230
  Charles A. Lindbergh, 338
  Sally Ride, 487
  Sir George Hubert Wilkins, 553

B

Balboa, Vasco Núñez de, 27-32
Ballard, Robert Duane, 33-38
Banks, Sir Joseph, 39-47
Bering, Vitus Jonassen, 48-54
Boone, Daniel, 55-61
Botany
  Sir Joseph Banks, 39
  Mungo Park, 405
Bougainville, Louis-Antoine de, 62-67
Bruce, James, 68-73
Bullock, Fanny. *See* Workman, Fanny Bullock
Burton, Sir Richard Francis, 74-80
Byrd, Richard E., 81-87

C

Cabeza de Vaca, Álvar Núñez, 88-93
Cabot, John, 94-100
Caboto, Juan. *See* Cabot, John
Canada. *See* North America (eastern), North America (northeastern), North America (northwestern)
Caribbean
  Vasco Núñez de Balboa, 27
  Christopher Columbus, 152
  Hernán Cortés, 179
  William Dampier, 203
  Francisco Pizarro, 437
  Juan Ponce de León, 451
  Amerigo Vespucci, 539
Carpini, Giovanni da Pian del, 101-107
Carson, Kit, 108-117
Carthaginian
  Hanno, 262
Cartier, Jacques, 118-124

Cavelier, René-Robert. *See* La Salle, Sieur de
Cavendish, Thomas, 125-130
Central America
    Vasco Núñez de Balboa, 27
    Christopher Columbus, 152
    Hernán Cortés, 179
    Francisco Pizarro, 437
    Hernando de Soto, 515
    Amerigo Vespucci, 539
Ch'en Yi. *See* Hsüan-tsang
Champlain, Samuel de, 131-137
Cheng Ho, 138-144
Chinese explorers
    Cheng Ho, 138
    Hsüan-tsang, 292
Clark, William, 329, 331, 333, 335, 337
Cochran, Jacqueline, 145-151
Colón, Cristóbal. *See* Columbus, Christopher
Colonization
    Stephen Fuller Austin, 19
    Sir Joseph Banks, 39
    Daniel Boone, 55
    Louis-Antoine de Bougainville, 62
    Jacques Cartier, 118
    Samuel de Champlain, 131
    Captain James Cook, 161
    Francisco Vásquez de Coronado, 170
    Hanno, 262
    Sieur de La Salle, 315
    Pedro Menéndez de Avilés, 377
    James Edward Oglethorpe, 396
    William Penn, 422
    Francisco Pizarro, 437
    Juan Ponce de Léon, 451
    Sir Walter Ralegh, 471
    John Smith, 509
    John Winthrop, 567
Columbus, Christopher, 152-160

Conservation
    Charles A. Lindbergh, 338
    John Muir, 383
Cook, Captain James, 161-169
Coronado, Francisco Vásquez de, 170-178
Cortés, Hernán, 179-187
Cousteau, Jacques-Yves, 188-195
Covilhã, Pêro da, 196-202

D

Dampier, William, 203-209
Danish explorers
    Vitus Jonassen Bering, 48
    Knud Johan Victor Rasmussen, 480
Davis, John, 210-215
Dias, Bartolomeu, 216-222
Diplomacy
    Sir Richard Francis Burton, 74
    Kit Carson, 108
    William Clark, 329
    Pêro da Covilhã, 196
    Sacagawea, 494
Drake, Sir Francis, 223-229

E

Earhart, Amelia, 230-237
East Indies
    Ibn Battutah, 309
Engineering
    Robert Duane Ballard, 33
    Jacques-Yves Cousteau, 188
    Charles A. Lindbergh, 338
    Sir George Hubert Wilkins, 553
English explorers
    Sir Joseph Banks, 39
    Sir Richard Francis Burton, 74
    Captain James Cook, 161
    William Dampier, 203
    John Davis, 210
    Sir Francis Drake, 223
    Martin Frobisher, 248
    Henry Hudson, 302

James Edward Oglethorpe, 396
William Penn, 422
Sir Walter Ralegh, 471
John Smith, 509
John Hanning Speke, 524
Henry Morton Stanley, 530
John Winthrop, 567
Ethnography
  James Bruce, 68
  Sir Richard Francis Burton, 74
  Thor Heyerdahl, 277
  Hsüan-tsang, 292
  Ibn Battutah, 309
  Marco Polo, 443
  John Wesley Powell, 457
  Knud Johan Victor Rasmussen, 480
  William of Rubrouck, 561
Europe
  Giovanni da Pian del Carpini, 101
  Fanny Bullock Workman, 574

F

Field exploration
  Roald Amundsen, 1
  Robert Duane Ballard, 33
  Captain James Cook, 161
  Jacques-Yves Cousteau, 188
  Thor Heyerdahl, 277
  Fridtjof Nansen, 389
  John Wesley Powell, 457
  Charles Wilkes, 547
  Sir George Hubert Wilkins, 553
  Fanny Bullock Workman, 574
Florida
  Pedro Menéndez de Avilés, 377
  Juan Ponce de Léon, 451
Frémont, John C., 238-247
French explorers
  Louis-Antoine de Bougainville, 62
  Jacques Cartier, 118
  Samuel de Champlain, 131

Jacques-Yves Cousteau, 188
Sieur de La Salle, 315
William of Rubrouck, 561
Frobisher, Martin, 248-254
Frontiersman
  Daniel Boone, 55
  Kit Carson, 108
  William Clark, 329
  Zebulon Montgomery Pike, 430
  Jedediah Strong Smith, 502

G

Gama, Vasco da, 255-261
Geography
  James Bruce, 68
  Sir Richard Francis Burton, 74
  Álvar Nuñez Cabeza de Vaca, 88
  John Cabot, 94
  Samuel de Champlain, 131
  William Clark, 329
  Captain James Cook, 161
  Pêro da Covilhã, 196
  John C. Frémont, 238
  Hanno, 262
  Prince Henry the Navigator, 269
  Hsüan-tsang, 292
  David Livingstone, 349
  Sir Alexander Mackenzie, 356
  Ferdinand Magellan, 363
  al-Mas'udi, 372
  Mungo Park, 405
  Zebulon Montgomery Pike, 430
  Marco Polo, 443
  John Wesley Powell, 457
  Pytheas, 465
  Knud Johan Victor Rasmussen, 480
  Sacagawea, 494
  John Smith, 509
  John Hanning Speke, 524
  Henry Morton Stanley, 530
  Amerigo Vespucci, 539
  William of Rubrouck, 561
  Fanny Bullock Workman, 574

INDEX

Government
  Stephen Fuller Austin, 19
  Cheng Ho, 138
  Francisco Vásquez de Coronado, 170
  John C. Frémont, 238
  Prince Henry the Navigator, 269
  William Penn, 422
  Sir Walter Ralegh, 471
  John Smith, 509
  Henry Morton Stanley, 530
  John Winthrop, 567
Greek explorer
  Pytheas, 465
Guides
  Kit Carson, 108
  Thomas Cavendish, 125
  Sacagawea, 494

H

Hanno, 262-268
Henry the Navigator, Prince, 269-276
Heyerdahl, Thor, 277-283
Hillary, Sir Edmund, 284-291
Historiography
  Giovanni da Pian del Carpini, 101
  al-Mas'udi, 372
Hsüan-tsang, 292-301
Hudson, Henry, 302-308
Humanitarianism
  Sir Edmund Hillary, 284
  Fridtjof Nansen, 389

I

Ibn Battutah, 309-314
India
  Pêro da Covilhã, 196
  Vasco da Gama, 255
Indian Ocean
  Bartolomeu Dias, 216
  Vasco da Gama, 255
  Thor Heyerdahl, 277

Italian explorers
  John Cabot, 94
  Giovanni da Pian del Carpini, 101
  Marco Polo, 443
  Amerigo Vespucci, 539
Italian/Spanish explorers
  Christopher Columbus, 152

L

La Salle, Sieur de, 315-321
Leif Eriksson, 322-328
Lewis, Meriwether, 329-337
Lindbergh, Charles A., 338-348
Literature
  Sir Richard Francis Burton, 74
  Hsüan-tsang, 292
  Ibn Battutah, 309
  al-Mas'udi, 372
  John Muir, 383
  Sir Walter Ralegh, 471
  John Smith, 509
  Henry Morton Stanley, 530
  Livingstone, David, 349-355

M

Ma San-po. *See* Cheng Ho
Mackenzie, Sir Alexander, 356-362
Magellan, Ferdinand, 363-371
Mariner
  Roald Amundsen, 1
  Vitus Jonassen Bering, 48
  Jacques Cartier, 118
  Cheng Ho, 138
  Christopher Columbus, 152
  Captain James Cook, 161
  William Dampier, 203
  John Davis, 210
  Bartolomeu Dias, 216
  Thor Heyerdahl, 277
  Henry Hudson, 302
  Leif Eriksson, 322
  Ferdinand Magellan, 363
  Martin Frobisher, 248

Pedro Menéndez de Avilés, 377
Charles Wilkes, 547
Mas'udi, al-, 372-376
Menéndez de Avilés, Pedro, 377-382
Mexico
   Álvar Nuñez Cabeza de Vaca, 88
Military
   Louis-Antoine de Bougainville, 62
   Richard E. Byrd, 81
   Kit Carson, 108
   Cheng Ho, 138
   William Clark, 329
   Jacqueline Cochran, 145
   Hernán Cortés, 179
   Pêro da Covilhã, 196
   Sir Francis Drake, 223
   John C. Frémont, 238
   Martin Frobisher, 248
   Vasco da Gama, 255
   Prince Henry the Navigator, 269
   Meriwether Lewis, 329
   Charles A. Lindbergh, 338
   Pedro Menéndez de Avilés, 377
   James Edward Oglethorpe, 396
   Robert Edwin Peary, 412
   Zebulon Montgomery Pike, 430
   Juan Ponce de León, 451
   Sir Walter Ralegh, 471
   Hernando de Soto, 515
   John Hanning Speke, 524
   Charles Wilkes, 547
   Sir George Hubert Wilkins, 553
Mountaineering
   Sir Edmund Hillary, 284
   Fanny Bullock Workman, 574
Muir, John, 383-388

N

Nansen, Fridtjof, 389-395
Natural science
   Sir Joseph Banks, 39
   Sir George Hubert Wilkins, 553

Navigation
   Vitus Jonassen Bering, 48
   Louis-Antoine de Bougainville, 62
   John Cabot, 94
   Jacques Cartier, 118
   Thomas Cavendish, 125
   Cheng Ho, 138
   Christopher Columbus, 152
   Captain James Cook, 161
   William Dampier, 203
   John Davis, 210
   Bartolomeu Dias, 216
   Sir Francis Drake, 223
   Vasco da Gama, 255
   Hanno, 262
   Thor Heyerdahl, 277
   Henry Hudson, 302
   Ferdinand Magellan, 363
   Pytheas, 465
   Sir Walter Ralegh, 471
   Amerigo Vespucci, 539
New Zealand
   Sir Joseph Banks, 39
   Sir Edmund Hillary, 284
North America
   Sir Walter Ralegh, 471
   Fanny Bullock Workman, 574
North America (central)
   Sieur de La Salle, 315
North America (eastern)
   William Penn, 422
   John Smith, 509
   John Winthrop, 567
North America (northeastern)
   Sir Joseph Banks, 39
   John Cabot, 94
   Jacques Cartier, 118
   Samuel de Champlain, 131
   Henry Hudson, 302
   Leif Eriksson, 322
North America (northern)
   Martin Frobisher, 248

INDEX

North America (northwestern)
  Sir Alexander Mackenzie, 356
North America (southeastern)
  Pedro Menéndez de Avilés, 377
  James Edward Oglethorpe, 396
  Hernando de Soto, 515
North America (southern)
  Álvar Nuñez Cabeza de Vaca, 88
North America (southwestern)
  Stephen Fuller Austin, 19
  Francisco Vásquez de Coronado, 170
  Zebulon Montgomery Pike, 430
North Atlantic
  Fridtjof Nansen, 389
  Pytheas, 465
Northwest Passage
  Roald Amundsen, 1
  John Davis, 210
  Martin Frobisher, 248
  Henry Hudson, 302
Norwegian explorers
  Roald Amundsen, 1
  Thor Heyerdahl, 277
  Leif Eriksson, 322
  Fridtjof Nansen, 389

O

Oceanography
  Robert Duane Ballard, 33
  Captain James Cook, 161
  Jacques-Yves Cousteau, 188
  Fridtjof Nansen, 389
Oglethorpe, James Edward, 396-404

P

Pacific Ocean
  Vasco Núñez de Balboa, 27
  Captain James Cook, 161
  William Dampier, 203
  Thor Heyerdahl, 277
  Ferdinand Magellan, 363
  Charles Wilkes, 547
Park, Mungo, 405-411
Peary, Robert Edwin, 412-421
Penn, William, 422-429
Pike, Zebulon Montgomery, 430-436
Piracy
  Thomas Cavendish, 125
  William Dampier, 203
  John Davis, 210
  Sir Francis Drake, 223
  Martin Frobisher, 248
  Sir Walter Ralegh, 471
Pizarro, Francisco, 437-442
Polo, Marco, 443-450
Ponce de Léon, Juan, 451-456
Portuguese explorers
  Bartolomeu Dias, 216
  Vasco da Gama, 255
  Prince Henry the Navigator, 269
Portuguese/Spanish explorers
  Ferdinand Magellan, 363
Powell, John Wesley, 457-464
Pytheas, 465-470

R

Ralegh, Sir Walter, 471-479
Raleigh, Sir Walter. *See* Ralegh, Sir Walter
Rasmussen, Knud Johan Victor, 480-486
Religion
  Giovanni da Pian del Carpini, 101
  Hsüan-tsang, 292
  David Livingstone, 349
  William Penn, 422
  William of Rubrouck, 561
  John Winthrop, 567
Ride, Sally, 487-493

Russia
  Vitus Jonassen Bering, 48
  Ibn Battutah, 309
  Ruysbroeck, Willem van. *See*
    William of Rubrouck

S

Sacagawea, 494-501
Sagagawea. *See* Sacagawea
Sakakawea. *See* Sacagawea
Scientific exploration
  Roald Amundsen, 1
  Robert Duane Ballard, 33
  Sir Joseph Banks, 39
  Captain James Cook, 161
  Jacques-Yves Cousteau, 188
  William Dampier, 203
  John C. Frémont, 238
  Thor Heyerdahl, 277
  Meriwether Lewis, 329
  John Muir, 383
  Fridtjof Nansen, 389
  John Wesley Powell, 457
  Pytheas, 465
  Sally Ride, 487
  Charles Wilkes, 547
  Sir George Hubert Wilkins, 553
Scottish explorers
  James Bruce, 68
  David Livingstone, 349
  Sir Alexander Mackenzie, 356
  John Muir, 383
  Mungo Park, 405
Smith, Jedediah Strong, 502-508
Smith, John, 509-514
Soto, Hernando de, 515-523
South America
  Sir Richard Francis Burton, 74
  John Cabot, 94
  Francisco Pizarro, 437
  Sir Walter Ralegh, 471
  Hernando de Soto, 515
  Amerigo Vespucci, 539

South Atlantic
  John Davis, 210
  Bartolomeu Dias, 216
  Vasco da Gama, 255
South Pacific
  Louis-Antoine de Bougainville, 62
Space
  Neil Armstrong, 10
  Sally Ride, 487
Spanish explorers
  Vasco Núñez de Balboa, 27
  Álvar Nuñez Cabeza de Vaca, 88
  Francisco Vásquez de Coronado, 170
  Hernán Cortés, 179
  Pedro Menéndez de Avilés, 377
  Francisco Pizarro, 437
  Juan Ponce de León, 451
  Hernando de Soto, 515
Spanish conquest
  Vasco Núñez de Balboa, 27
  Álvar Nuñez Cabeza de Vaca, 88
  Francisco Vásquez de Coronado, 170
  Hernán Cortés, 179
  Pedro Menéndez de Avilés, 377
  Francisco Pizarro, 437
  Juan Ponce de León, 451
  Hernando de Soto, 515
Speke, John Hanning, 524-529
Stanley, Henry Morton, 530-538
Statesmanship
  Stephen Fuller Austin, 19
  Fridtjof Nansen, 389
Subarctic. *See* Arctic

T

T'ang San-tsang. *See* Hsüan-tsang
Topography
  Álvar Nuñez Cabeza de Vaca, 88
  Zebulon Montgomery Pike, 430
  John Wesley Powell, 457
Tripitaka. *See* Hsüan-tsang

U

Undersea
  Robert Duane Ballard, 33
  Jacques-Yves Cousteau, 188

V

Vespucci, Amerigo, 539-546

W

Wilkes, Charles, 547-552
Wilkins, Sir George Hubert, 553-560
William of Rubrouck, 561-566
Winthrop, John, 567-573
Workman, Fanny Bullock, 574-580

World circumnavigation
  Sir Joseph Banks, 39
  Louis-Antoine de Bougainville, 62
  Thomas Cavendish, 125
  Captain James Cook, 161
  William Dampier, 203
  Sir Francis Drake, 223
  Ferdinand Magellan, 363
  Charles Wilkes, 547

Z

Zoology
  Jacques-Yves Cousteau, 188
  Fridtjof Nansen, 389